AF576000

Quality and Safety of Meat Products

Quality and Safety of Meat Products

Editors

Begoña Panea
Guillermo Ripoll

MDPI • Basel • Beijing • Wuhan • Barcelona • Belgrade • Manchester • Tokyo • Cluj • Tianjin

Editors
Begoña Panea
CITA-Zaragoza University
Spain

Guillermo Ripoll
Unidad de Producción y Sanidad Animal
Spain

Editorial Office
MDPI
St. Alban-Anlage 66
4052 Basel, Switzerland

This is a reprint of articles from the Special Issue published online in the open access journal *Foods* (ISSN 2304-8158) (available at: https://www.mdpi.com/journal/foods/special_issues/Quality_Safety_Meat).

For citation purposes, cite each article independently as indicated on the article page online and as indicated below:

LastName, A.A.; LastName, B.B.; LastName, C.C. Article Title. *Journal Name* **Year**, *Article Number*, Page Range.

ISBN 978-3-03943-372-8 (Hbk)
ISBN 978-3-03943-373-5 (PDF)

Cover image courtesy of Guillermo Ripoll.

Contents

About the Editors vii

Begoña Panea and Guillermo Ripoll
Quality and Safety of Meat Products
Reprinted from: *Foods* **2020**, *9*, 803, doi:10.3390/foods9060803 1

Ana Guerrero, Carlos Sañudo, María del Mar Campo, Jose Luis Olleta, Erica Muela, Rosa M. G. Macedo and Francisco A. F. Macedo
Consumer Acceptability of Dry Cured Meat from Cull Ewes Reared with Different Linseed Supplementation Levels and Feeding Durations
Reprinted from: *Foods* **2018**, *7*, 89, doi:10.3390/foods7060089 5

Guillermo Ripoll, Margalida Joy and Begoña Panea
Consumer Perception of the Quality of Lamb and Lamb Confit
Reprinted from: *Foods* **2018**, *7*, 80, doi:10.3390/foods7050080 15

Martha Y. Leal-Ramos, Alma D. Alarcón-Rojo, Néstor Gutiérrez-Méndez, Hugo Mújica-Paz, Felipe Rodríguez-Almeida and Armando Quintero-Ramos
Improving Cull Cow Meat Quality Using Vacuum Impregnation
Reprinted from: *Foods* **2018**, *7*, 74, doi:10.3390/foods7050074 29

Andiara Gonçalves-Tenório, Beatriz Nunes Silva, Vânia Rodrigues, Vasco Cadavez and Ursula Gonzales-Barron
Prevalence of Pathogens in Poultry Meat: A Meta-Analysis of European Published Surveys
Reprinted from: *Foods* **2018**, *7*, 69, doi:10.3390/foods7050069 41

Djamel Djenane and Pedro Roncalés
Carbon Monoxide in Meat and Fish Packaging: Advantages and Limits
Reprinted from: *Foods* **2018**, *7*, 12, doi:10.3390/foods7020012 57

Ioannis Konstantinos Karabagias
Volatile Profile of Raw Lamb Meat Stored at 4 ± 1 °C: The Potential of Specific Aldehyde Ratios as Indicators of Lamb Meat Quality
Reprinted from: *Foods* **2018**, *7*, 40, doi:10.3390/foods7030040 91

KaWang Li, Amanda Gipe McKeith, Cangliang Shen and Russell McKeith
A Comparison Study of Quality Attributes of Ground Beef and Veal Patties and Thermal Inactivation of *Escherichia coli* O157:H7 after Double Pan-Broiling Under Dynamic Conditions
Reprinted from: *Foods* **2018**, *7*, 1, doi:10.3390/foods7010001 103

Peter J. Bechtel, John M. Bland, Kristin Woods, Jeanne M. Lea, Suzanne S. Brashear, Stephen M. Boue, Kim W. Daigle and Karen L. Bett-Garber
Effect of Par Frying on Composition and Texture of Breaded and Battered Catfish
Reprinted from: *Foods* **2018**, *7*, 46, doi:10.3390/foods7040046 117

Brianne A. Altmann, Carmen Neumann, Susanne Velten, Frank Liebert and Daniel Mörlein
Meat Quality Derived from High Inclusion of a Micro-Alga or Insect Meal as an Alternative Protein Source in Poultry Diets: A Pilot Study
Reprinted from: *Foods* **2018**, *7*, 34, doi:10.3390/foods7030034 129

About the Editors

Begoña Panea graduated in Veterinary Medicine, specializing in Bromatology, Health and Food Technology at the University of Zaragoza (1989), and obtained her PhD in Veterinary Medicine at the University of Zaragoza (2002). She is currently a researcher at the Centro de Investigación y Tecnología Agroalimentaria de Aragón (Zaragoza, Spain). She has experience in the area of Animal Production and Meat and Meat Products Science, acting mainly on the following themes: production factors that affect the quality of the carcass and meat, product classification and diversification, useful life of meat and meat products, meat products, the sensorial analysis of food, and studies with consumers.

Guillermo Ripoll (Ph.D) is an agricultural engineer and master in Agrifood Biotecnology. He is currently a technical researcher at the Centro de Investigación y Tecnología Agroalimentaria de Aragón (Zaragoza, Spain). The research of the team focuses on pre- and post-slaughter factors affecting meat shelf life and quality. His preferred topic is meat color and he is especially interested in the consumers' perception of meat color. Sometimes he also deals with NIR spectroscospy.

Editorial

Quality and Safety of Meat Products

Begoña Panea [1,2,*] and Guillermo Ripoll [1,2]

1 Centro de Investigación y Tecnología Agroalimentaria de Aragón (CITA), Unidad de Producción y Sanidad Animal, Avda. Montañana 930, 50059 Zaragoza, Spain; gripoll@aragon.es
2 Instituto Agroalimentario de Aragón–IA2 (CITA)-Universidad de Zaragoza, 50013 Zaragoza, Spain
* Correspondence: bpanea@cita-aragon.es; Tel.: +34-976-716-443

Received: 1 June 2020; Accepted: 15 June 2020; Published: 18 June 2020

Abstract: Food safety is a major problem around the world, both regarding human suffering and with respect to economic costs. Scientific advances have increased our knowledge surrounding the nutritional characteristics of foods and their effects on health. This means that a large proportion of consumers are much more conscious with respect to what they eat and their demands for quality food. Food quality is a complex term that includes, in addition to safety, other intrinsic characteristics, such as appearance, color, texture and flavor, and also extrinsic characteristics, such as perception or involvement.

Keywords: poultry; beef; lamb; carbon monoxide; volatile compounds; packaging; enhanced meat; sensory analysis; Campylobacter; *Escherichia coli O157:H7*; Hermetia illucens; Listeria monocytogenes; *Staphylococcus aureus*; Salmonella

---

Food-borne diseases are a main problem in the meat industry [1]. The study of pathogens present in meat and meat products is crucial for the industry, for sanitary administration and to generate consumer trust. *Escherichia coli O157:H7* is one of the most important and studied pathogens present in fresh meat and it has been considered a contaminant of raw, non-intact beef products since 1999 [2]. A recent survey showed that 40–58% of US consumers ordered beefsteaks at medium rare (60–62.8 °C) to rare (54.4–57.2 °C), which could potentially put consumers at a high risk from *E. coli O157:H7* [3]. Some techniques should be implemented to reduce this risk. Li, et al. [4] conducted a study on coarse ground beef and veal patties, which aimed to investigate the quality variances, including color variation, during aerobic storage and cooking, as well as to evaluate the thermal inactivation of *E. coli O157:H7*. They hypothesize that a higher internal temperature with a longer rest time will increase the inactivation of *E. coli O157:H7* in beef and veal patties. The results showed that *Escherichia coli O157:H7* was more sensitive to heat in veal compared to beef, with shorter D-value. Cooking to 71.1 and 76 °C reduced *E. coli O157:H7* by >6 log CFU/g, regardless of rest time. Cooking to 55 °C and 62.5 °C with a 3.5 min rest achieved an additional 1–3 log CFU/g reduction compared to the 0.5 min rest. These results should be useful for developing a risk assessment of non-intact beef and veal products.

Behind this well-known pathogen, an additional problem is the emergence of other pathogens. Among this group of infectious bacteria, *Salmonella* spp., *Listeria monocytogenes*, *enterotoxigenic*, *Staphylococcus aureus* and *Campylobacter* spp. are the main contaminants in food due to their high occurrence worldwide and being major causes of gastroenteritis in humans [5,6]. Goncalves-Tenorio, et al. [7] published a meta-analysis, which summarized the levels of incidence of *Salmonella* spp., *Listeria monocytogenes*, *Staphylococcus aureus* and *Campylobacter* spp. in poultry meat commercialized in Europe. The results suggest that *S. aureus* is the main pathogen detected in poultry meat (38.5%; 95%CI: 25.4–53.4), followed by *Campylobacter* spp. (33.3%; 95% CI: 22.3–46.4%), while *L. monocytogenes* and *Salmonella* spp. present lower prevalence (19.3%; 95% CI: 14.4–25.3% and 7.10%; 95% CI: 4.60–10.8%, respectively). Despite the differences in prevalence, all of the pathogens were found in chicken and other poultry meats, at both the end-processing step and at retail level, in packed and unpacked products and in several meat cutting types. Prevalence data on cold preservation products also revealed that chilling and freezing can reduce the proliferation of pathogens but do not inactivate them.

Because cold application is not enough to prevent meat spoilage, it should be complemented with other technologies, such as packaging. During the last few decades, the modified atmosphere packaging (MAP) of foods has been a promising area of research, but much remains to be known regarding the use of unconventional gases, such carbon monoxide (CO). The use of CO for meat and seafood packaging is not allowed in most countries due to its potential toxic effect, and its use is controversial in some countries. Djenane and Roncalés [8] undertake a review to present the most comprehensive and current overview of the widely available, scattered information about the use of CO in the preservation of muscle foods. The conclusions shown that the use of CO in fresh meat packaging gives promising results due to its positive effects on overall meat quality, especially on color. The results stated that the risk of CO toxicity from the packaging process or from consumption of CO-treated meats is negligible and, additionally, CO is not present in the pack during storage. Some recommendations and future prospects addressed to food industries, consumers and regulators were pointed as "best practices".

However, the packaging causes changes in meat quality, especially in color and flavor and, since these attributes are used by consumers to evaluate meat freshness, these changes due to the packaging must to be investigated [9]. A study was carried out by Karabagias [10] to evaluate the volatile profile of raw lamb meat during storage under refrigeration and to evaluate the aroma evolution of raw lamb packaged in a multi-layer coating film and stored at 4 ± 1 °C, as well as to investigate whether specific aldehyde ratios could serve as markers of lamb-meat freshness and degree of oxidation. Volatile compounds were determined using headspace solid phase microextraction coupled with gas chromatography/mass spectrometry. The results showed that volatile compound content increased during storage time and that the evolution of aldehydes during storage recorded a positive Pearson's correlation (r) ($p < 0.05$) with the degree of oxidation (mg malonic dialdehyde per kg of lamb meat). In addition, a perfect Pearson's correlation (r = 1) was obtained for the ratio hexanal to nonanal and, therefore, this ratio was proposed as an indicator of lamb meat freshness and overall quality.

Besides safety, another important issue for farmers and retailers is the cost-effectiveness. On the one hand, the main cost for farmers, especially those that reared monogastric animals, is feedstuffs, which in turn is dependent on soybean global markets and prices. Then, other protein sources have been investigated and algae and insects seem to be good substitutes. For this purpose, Altmann, et al. [11] raised 132 Ross 308 male birds on amino acid balanced diets, where 50% of the soy-based protein was substituted by either spirulina powder (*Arthrospira platensis*) or *Hermetia illucens* partially defatted larval meal, in starter and grower diets. Slaughterhouse parameters and meat physico-chemical and sensory properties were investigated. The results showed that meat quality could be improved when spirulina replaced 50% of the soy protein in broiler diets, although this substitution resulted in a dark red-yellow meat color. Besides, the substitution with *Hermetia* larval meal resulted in a product that did not differ from the standard fed control group, with the exception that the breast filet had a more intense flavor that decreased over storage time. Then, it was concluded that spirulina and *Hermetia* meal have the potential to replace soybean meal in broiler diets without deteriorating meat quality.

To modify feedstuff also allows us to obtain new products and, if these products are elaborated using the less valorized carcass joints, we have a feasible strategy to improve the incomes of farmers. This idea is the basis of the experiment carried out by Guerrero, et al. [12]. The paper assesses consumer acceptability of a cured product ("Cecina") elaborated with cull ewes meat finished with different levels of linseed (5, 10 or 15%) for different periods before slaughtering (30, 50 or 70 days). The results show that linseed supplementation was identified as the most important factor for sensorial attributes ($p < 0.01$), with the preferred "Cecina" being that with 5% and 10% supplementation.

Another major threat to the sector is the constant decrease in meat consumption over the world [13]. The industry answered by implementing two main strategies: product enhancement and the development of new products.

Several approaches are possible to enhance a product [14,15]. A commonly used alternative is the addition of substances that improve the physical properties of the food (e.g., moisture enhancers). This technology improves tenderness, juiciness, flavor, and consistency of whole-muscle meat products, especially in those of reduced eating quality [16,17]. Unfortunately, with the current industry for brine injection, brine is not uniformly distributed throughout the meat. Vacuum impregnation allows for the direct insertion of an external solution into a product through its pores in a fast, controlled and uniform way without destroying its original structure [18]. Leal-Ramos, et al. [19] conducted an experiment to vary the pressure, (20.3, 71.1 kPa) and time, (0.5, 2.0, 4.0 h) of impregnation. The results showed that both the vacuum and atmospheric pressures generated a positive impregnation and deformation. The highest values of impregnation (10.5%) and deformation (9.3%) were obtained at a pressure of 71.1 kPa and a time of 4.0 h. The sample effective porosity exhibited a significant interaction ($p < 0.01$) between pressure and time and the highest value (14.0%) was achieved at a pressure of 20.3 kPa and a time of 4.0 h, whereas the most extended distension of meat fibers (98 μm) was observed at the highest levels of pressure and time. These results indicate that meat from mature cows can undergo a vacuum-wetting process successfully, with an IS of sodium chloride to improve its quality.

The enhancement can also be afforded regarding nutritional value and, in this approach, the search for low-calorie products has been widely studied and developed in several kind of products [20]. Bechtel, et al. [21] aimed to make a battered catfish product that could be baked with a lower percentage of oil-based calories than the equivalent par fried products. Then, the effect of different batters (rice, corn, and wheat) was examined and the effect of par frying on the composition and texture properties of baked catfish. The results found that the lipid contents of the par fried treatments were significantly higher for both corn and wheat batters than for non-par fried treatments. Sensory analysis indicated that the texture of the coatings in the par fried treatments were significantly greater for hardness attributes. In addition, fillet flakiness was significantly greater in the par fried treatments and the corn-based batters had moister fillet strips compared to the wheat flour batters.

Finally, interest in convenience products has increased in the last few years because the patterns of food and meat consumption are constantly changing, not only due to socioeconomic and cultural trends that affect the whole society, but also to the specific lifestyles of consumer groups [22]. Ripoll, et al. [23] carried out a study to identify the profiles of lamb meat consumers according to their orientation toward convenience, also analyzing their socioeconomic characteristics and their preferences regarding the intrinsic and extrinsic quality signals of lamb meat and their willingness to pay for lamb confit. Four types of consumers were differentiated according to their lifestyles related to lamb consumption: "Gourmet", "Disinterested", "Conservative" and "Basic". The Gourmet group has characteristics that make it especially interesting to market a product, such as lamb confit, and consequently, marketing strategy should to be focused to this niche market.

Author Contributions: The authors have made a substantial, direct and intellectual contribution to the work and approved it for publication. All authors have read and agreed to the published version of the manuscript.

Funding: This research received no external funding.

Conflicts of Interest: The authors declare no conflict of interest.

References

1. Fegan, N.; Jenson, I. The role of meat in foodborne disease: Is there a coming revolution in risk assessment and management? *Meat Sci.* **2018**, *144*, 22–29. [CrossRef]
2. USDA. Beef products contaminated with *Escherichia coli* O157: H7. *Fed Regist.* **1999**, *64*, 2803–2804.
3. Schmidt, T.; Keene, M.; Lorenzen, C. Improving consumer satisfaction of beef through the use of thermometers and consumer education by wait staff. *J. Food Sci.* **2002**, *67*, 3190–3193.

4. Li, K.; McKeith, A.G.; Shen, C.; McKeith, R. A comparison study of quality attributes of ground beef and veal patties and thermal inactivation of Escherichia coli O157: H7 after double pan-broiling under dynamic conditions. *Foods* **2018**, *7*, 1.
5. Van Nierop, W.; Duse, A.; Marais, E.; Aithma, N.; Thothobolo, N.; Kassel, M.; Stewart, R.; Potgieter, A.; Fernandes, B.; Galpin, J. Contamination of chicken carcasses in Gauteng, South Africa, by Salmonella, Listeria monocytogenes and Campylobacter. *Int. J. Food Microbiol.* **2005**, *99*, 1–6.
6. Hennekinne, J.-A.; De Buyser, M.-L.; Dragacci, S. Staphylococcus aureus and its food poisoning toxins: Characterization and outbreak investigation. *FEMS Microbiol. Rev.* **2012**, *36*, 815–836.
7. Goncalves-Tenorio, A.; Silva, B.N.; Rodrigues, V.; Cadavez, V.; Gonzales-Barron, U. Prevalence of Pathogens in Poultry Meat: A Meta-Analysis of European Published Surveys. *Foods* **2018**, *7*. [CrossRef]
8. Djenane, D.; Roncalés, P. Carbon monoxide in meat and fish packaging: Advantages and limits. *Foods* **2018**, *7*, 12.
9. Realini, C.E.; Marcos, B. Active and intelligent packaging systems for a modern society. *Meat Sci.* **2014**, *98*, 404–419. [CrossRef]
10. Karabagias, I.K. Volatile profile of raw lamb meat stored at 4±1 C: The potential of specific aldehyde ratios as indicators of lamb meat quality. *Foods* **2018**, *7*, 40.
11. Altmann, B.A.; Neumann, C.; Velten, S.; Liebert, F.; Mörlein, D. Meat quality derived from high inclusion of a micro-alga or insect meal as an alternative protein source in poultry diets: A pilot study. *Foods* **2018**, *7*, 34.
12. Guerrero, A.; Sañudo, C.; Campo, M.D.M.; Olleta, J.L.; Muela, E.; Macedo, R.M.; Macedo, F.A. Consumer Acceptability of Dry Cured Meat from Cull Ewes Reared with Different Linseed Supplementation Levels and Feeding Durations. *Foods* **2018**, *7*, 89.
13. Rodríguez-Serrano, T.M.; Panea, B.; Alcalde, M.J. A European vision for the small ruminant sector. Promotion of meat consumption campaigns. *Small Rumin. Res.* **2016**, *142*, 3–5. [CrossRef]
14. Jiménez-Colmenero, F. Potential applications of multiple emulsions in the development of healthy and functional foods. *Food Res. Int.* **2013**, *52*, 64–74. [CrossRef]
15. Siro, I.; Kapolna, E.; Kapolna, B.; Lugasi, A. Functional food. Product development, marketing and consumer acceptance–a review. *Appetite* **2008**, *51*, 456–467. [CrossRef]
16. Feiner, G. *Meat Products Handbook: Practical Science and Technology*; Elsevier: Amsterdam, The Netherlands, 2006.
17. Vihavainen, E.J.; Björkroth, K.J. Spoilage of value-added, high-oxygen modified-atmosphere packaged raw beef steaks by Leuconostoc gasicomitatum and Leuconostoc gelidum. *Int. J. Food Microbiol.* **2007**, *119*, 340–345.
18. Salvatori, D.; Andres, A.; Chiralt, A.; Fito, P. The response of some properties of fruits to vacuum impregnation. *J. Food Process Eng.* **1998**, *21*, 59–73.
19. Leal-Ramos, M.Y.; Alarcón-Rojo, A.D.; Gutiérrez-Méndez, N.; Mújica-Paz, H.; Rodríguez-Almeida, F.; Quintero-Ramos, A. Improving Cull Cow Meat Quality Using Vacuum Impregnation. *Foods* **2018**, *7*, 74.
20. Ospina, E.J.; Sierra, C.A.; Ochoa, O.; Perez-Alvarez, J.A.; Fernandez-Lopez, J. Substitution of saturated fat in processed meat products: A review. *Crit. Rev. Food Sci. Nutr.* **2012**, *52*, 113–122. [CrossRef]
21. Bechtel, P.J.; Bland, J.M.; Woods, K.; Lea, J.M.; Brashear, S.S.; Boue, S.M.; Daigle, K.W.; Bett-Garber, K.L. Effect of Par Frying on Composition and Texture of Breaded and Battered Catfish. *Foods* **2018**, *7*, 46.
22. Bernués, A.; Ripoll, G.; Panea, B. Consumer segmentation based on convenience orientation and attitudes towards quality attributes of lamb meat. *Food Q. Prefer.* **2012**, *26*, 211–220. [CrossRef]
23. Ripoll, G.; Joy, M.; Panea, B. Consumer Perception of the Quality of Lamb and Lamb Confit. *Foods* **2018**, *7*, 80. [CrossRef]

Article

Consumer Acceptability of Dry Cured Meat from Cull Ewes Reared with Different Linseed Supplementation Levels and Feeding Durations

Ana Guerrero [1,*], Carlos Sañudo [1], María del Mar Campo [1], Jose Luis Olleta [1], Erica Muela [1], Rosa M. G. Macedo [1,2] and Francisco A. F. Macedo [1,2]

1 Department of Animal Production and Food Science, Instituto Agroalimentario (IA2), Universidad de Zaragoza—CITA, C/Miguel Servet, 177, 50013 Zaragoza, Spain; csanudo@unizar.es (C.S.); marimar@unizar.es (M.d.M.C.); olleta@unizar.es (J.L.O.); ericamola@hotmail.com (E.M.); rmgmacedo@uem.br (R.M.G.M.); fafmacedo@uem.br (F.A.F.M.)

2 Department of Animal Science, Universidade Federal, de Sergipe, Cidade Universitária, 49100-000 São Cristovão, Brazil

* Correspondence: aguerre@unizar.es; Tel.:+34-876-55-41-48

Received: 22 May 2018; Accepted: 1 June 2018; Published: 11 June 2018

Abstract: Dry cured meat—'cecina'—is a traditional, although not well-known, dry product that could add value to cull ewes. Because of this, the aim of the study was to assess consumer acceptability of 'cecina' from cull ewes finished with different levels of linseed (5, 10 or 15%) for different periods before slaughtering (30, 50 or 70 days). One hundred and fifty consumers evaluated colour acceptability, fatness and odour, flavour and overall acceptability of 'cecina' from those 9 treatments. Additionally, habits of consumption of cured products and preferences for different species and willingness to pay for 'cecina' were investigated. Linseed supplementation was identified as the most important factor for sensorial attributes ($p < 0.01$), with the preferred 'cecina' being that with 5% and 10% supplementation. Feeding duration only modified the fatness acceptability ($p < 0.01$). 'Cecina' from small ruminants is a product consumed occasionally by the majority of participants; however, it presented an adequate overall acceptability. Consequently, elaborating 'cecina' would be a feasible strategy to improve the income of farmers.

Keywords: cecina; ovine; sensory quality; traditional meat products

---

1. Introduction

'Cecina' is a traditional meat product that can be elaborated from several species after salting, drying and, occasionally, smoking different pieces of meat, mainly back leg and sirloin [1]. This product is particularly appreciated and consumed in some countries of the Mediterranean area, but it has equivalents in many other areas of the world [2].

Depending on the region, species, joint and specific local variations on the production and fabrication processes, different products can be found with different denominations. Spanish 'cecina' resembles South African 'biltong', South American 'charqui', Italian 'bresaola' and Turkish 'pastirma' [1]. All of these have several common characteristics, such as their distinctive flavour (one of the key attributes for the consumer) and its local production following traditional processes.

In many Mediterranean countries, as currently happens in Spain, the ovine meat sector is important in the maintenance and sustainability of rural areas [3]. Usually, fresh meat from cull ewes has little economic value. However, dried and cured meat from those types of animals increases their commercial value [4], although dry-cured meat products from small ruminant are less frequent than from other species (beef, pork) [5]. Therefore, the transformation to 'cecina', which presumably presents a good overall acceptability according to previous studies [2], would have a positive economic outcome for breeders.

However, cull animals, which represent up to 40% of ewes in the flock [6], could need a slight improvement in their body score and conformation before slaughtering in order to improve their yield and carcass quality, and maybe the meat characteristics. With these aims, ref. [7] studied the effect of the addition of different levels of linseed in the diet during different fattening periods on meat quality. Linseed was selected as an alternative ingredient to use in the finishing diets of those animals due to its energetic content, its nutritional profile, which is rich in n-3 polyunsaturated fatty acids, and its demonstrated ability to modify meat fatty acid composition, resulting in a healthier profile [8,9].

In addition, for obtaining market success of processed sheep meat, it is essential to know consumer sensory and hedonic perception [4]. Sensory attributes are one of the main characteristics underlying consumers' overall liking, but other variables, such as familiarity with the product, socio-cognitive segmentation or genetic origins of the consumers would also be important topics to be considered and discussed [10] in order to better understand consumer acceptability of the tested product.

Consequently, the current work examines the influence of the different levels of linseed supplementation and feeding duration in cull ewes on the consumer acceptability of the dry cured sheep meat 'cecina'.

2. Material and Methods

2.1. Animals and Carcass Quality

Seventy-two ewes of the Rasa Aragonesa breed, a local medium wool breed, rustic type from the North East (Aragón) of Spain (for more information see [11]) were randomly selected from the cooperative Grupo Pastores® from 5 commercial flocks, with all animals presenting similar characteristics of reproductive life, age (older than 6 years) and rearing conditions (semi-intensive systems). After their selection, cull ewes were transported to the facilities of the Veterinary Faculty of Zaragoza, where they were intensively finished based on concentrates and cereal straw. The experimental design was composed of 9 groups with 8 animals per group. The different experimental diets evaluated were isonitrogenous and isoenergetics.

A 3 × 3 factorial design was tested, which comprised 3 different percentages of *Linum usitatissimum* (linseed), which was supplied as whole seeds on the concentrate diet (linseed supplementation level—LSL): 5%, 10% or 15%. Also, 3 different finishing duration periods (FD) where evaluated: 30, 50 or 70 days.

Productive parameters for growing and animal performance, such as live weight, body condition score, average daily gain, were compiled throughout the experiment, and were compiled in [7]. When the corresponding finishing periods were reached, animals were transported (<3 h) to an EU-licensed abattoir to be slaughtered after a resting period of 15–20 h. Then, carcasses were kept at between 2 and 4 °C overnight. Twenty-four hours post-slaughter, the left side of the carcass was transported refrigerated to the Meat Quality Laboratory of the Veterinary Faculty of Zaragoza, where the hind limb was excised and used to elaborate the 'cecina' to be evaluated by consumers.

2.2. Curing Procedure

Each hind limb, after being excised, was vacuum packed and sent with 4 days of ageing to the food industry, where they were processed (Secadero Sierra Maestrazgo®, Castellote, Teruel, Spain) in order to obtain 'cecina' (cure meat) with the same methodology described in [2].

The curing procedure consisted of piling and covering the surface of the hind legs with marine salt in a cold room, inverting the location of the pieces periodically to guarantee the homogenous penetration of a convenient amount of salt on the pieces. Later, salt residues on the surface of the meat pieces were eliminated by washing them with cold water, and for salt equilibration, pieces were hung in a cold room (5 °C; 80% humidity) for a period of between 45 and 50 days. Finally, drying was conducted by hanging the meat pieces in natural conditions over winter (2 months, room temperature of 15 °C, relative humidity of 85–90%). During the drying process, biochemical reactions occur, giving the characteristic and typical flavours and textures associated with 'cecina' by ripening [12]. Once the technician had certified that each 'cecina' batch had finished its curing process and it was ready for consumption, the legs were retired, deboned, vacuum packed, and kept at 4 °C until consumer analysis.

2.3. Consumer Test

Home-based tests [13] were used to determinate the acceptability of 'cecina'. In the trial, 150 local consumers were selected by gender and age according to the Spanish national profile [14] were involved. 54% of participants were women and 46% men. With regard to the age interval, 22.7% of consumers were between 18 and 34 years old; 47.3% were between 35 and 50 years old, and 30.0% were between 51 and 80 years old. All participants were of Spanish origin and were residents in Zaragoza. Consumers came from 46 different families, with 1 to 4 members each one.

Participants filled in a survey including closed questions with multiple choices, in order to obtain the frequency of consumption of cured products, knowledge about species from which it is possible to obtain 'cecina', and willingness to pay for those products.

The deboned 'cecina' was sliced with a continuous blade machine (Braher®mod. DSB28, Braher International, San Sebastian, Spain) into 1.5-mm-thick slices and packaged in three-digit-coded aluminium foil and vacuum packed together with all of the samples that each consumer was to evaluate.

Each consumer assessed 9 samples, one from each dietary treatment studied (3 linseed supplementation level (LSL) × 3 finishing duration periods (FD)). Consumers were asked to eat unsalted toasted bread and rinse their mouth with water before evaluating each sample. For each sample, consumers evaluated the acceptability of the following attributes: colour, fatness, odour, flavour and overall, by using a hedonic 9-point scale ranging from 1 (dislike extremely) to 9 (like extremely). The neutral central point (neither like nor dislike) had been deleted in an attempt to force the consumer to make a decision in a positive or negative way [15].

2.4. Statistical Analysis

Analysis of variance was performed using the General Linear Model (GLM) procedure of the SPSS statistical package (v.19.0, IBM Corp., Armonk, NY, USA). For consumer preference, linseed supplementation level (LSL), feeding duration (FD) and their interactions were considered as fixed effects, and consumer as a random effect.

The model was as follows:

$$Y_{ijk} = \mu + LSL_i + FD_j + A_k + (LSL_i \times FD_j) + e_{ijk}$$

where Y_{ijk} is the dependent variable; μ is the population average; LSL_i is the fixed effect of linseed supplementation level (5, 10 or 15%); FD_j is the fixed effect of supplementation period (30, 50 or 70 days); A_k is the random effect of the consumer; $(LSL_i \times FD_j)$ is the interaction effect of linseed supplementation and period; and e_{ijk} is the random error. The mean and standard error of the mean (SEM) were calculated. Differences between means were evaluated using Duncan's Multiple Range Test, ($p \leq 0.05$). A hierarchical cluster analysis was performed on overall acceptability using XLSTAT 16.5 in order to identify similarities among consumers.

3. Results and Discussion

3.1. Characterisation of Consumer Sample

The frequencies of consumption of different meat products by the participants of the study are compiled in Table 1. It can be observed that 'cecina' is consumed very sporadically by the majority of participants (82.8%), presenting lower frequencies of monthly and weekly consumption (8.3% and 1.4%, respectively). However, consumption of other dry cured products derived from pork, especially ham, is higher. Those results agree with other studies [16], in which 26% of participants consumed ham daily, and 67% several days per week. In Spain, there is a high culture of ham consumption (2.41 kg per capita, which represents about a quarter of the total meat products consumed [17], and a broad knowledge of its characteristics and brands [16]. However, in other European countries consumption of ham is less than once per week (Belgium, Denmark, Poland, Greece or Germany) [18], or very occasional in some other countries with high meat consumption such as Brazil [19].

Table 1. Frequency of consumption of different meat products (percentage; n = 145 consumers).

	Almost Every Day	2–3 t/w	Once/w	Once/15 d	Once/m	Ocassionally	Not Answer
Cecina	0.0	1.4	1.4	1.4	8.3	82.8	4.8
Ham	10.3	50.3	24.1	9.0	4.8	1.4	0.0
Cured loin	2.1	9.0	26.2	18.6	21.4	19.3	3.4
Chorizo	3.4	20.0	28.3	17.9	12.4	17.2	0.7
Fuet	2.1	16.6	20.0	13.1	15.9	26.9	5.5
Others	6.2	4.8	11.7	6.9	6.2	22.8	41.4

t: times; w: week; d: day; m: month.

Related to willingness to pay for 'cecina' (Table 2), participants were informed that the average price of cured ham in the market was 15 €/kg, and different possible prices for 'cecina' were subsequently presented (ranging from 10.5 €/kg to 19.5 €/kg). 15.2% of participants would pay the same price for 'cecina' as for ham (15 €/kg), 27.6% would pay a higher price for the product 'cecina' (16.5–18 €/kg), with a low percentage (2.1%) being willing to pay up to 19.5 €/kg. Approximately half of the answers indicated that consumers would pay a price lower for 'cecina' than the price of ham. Consumer education into "gourmet gastronomy" would be an alternative to improve success in the promotion and choice of a not-well-known product [19], such as 'cecina'.

Table 2. Price ranges that consumers would be willing to pay for 'cecina' (percentage; n = 145 consumers).

Price Ranges	Percentage of Consumers
10.5 €/kg	24.1
12.0 €/kg	17.2
13.5 €/kg	10.3
15.0 €/kg	15.2
16.5 €/kg	11.7
18.0 €/kg	15.9
19.5 €/kg	2.1
Not answer	3.4

Another question in the survey explored the knowledge related to the varieties of 'cecina' available. Consumers were asked to mark the species that they thought had 'cecina' in the market, with the possibility of marking multiple answers. The most popular species for 'cecina' that the majority of consumers knew were: deer (81.4%), cow (71.7%), wild pig (69.0%) and horse (62.8%). 'Cecina' from small ruminants was less popular; only 37.2% of consumers thought that 'cecina' from goat existed, and 31.0% of consumers thought the same for sheep. 'Cecina' from pig, duck or rabbit were little known by consumers (37.2%, 22.1% and 4.8%, respectively). In this sense, in Spain, the only 'cecina' with Protected Geographic Indication (PGI) is from beef ('Cecina de León').

These results can help to understand the answers compiled in Table 3 about the species that consumers preferred for 'cecina', because consumers usually prefer already known and experienced products, and in Spain the most popular 'cecina' product comes from beef. In any case, there is a small niche of consumers (11.7%) who chose 'cecina' from sheep as one of the three preferred first-choice options. Data from several sensory studies [10,20] has shown that familiar foods, and exposure to new foods, contribute to reducing the possible initial neophobia towards novel or unfamiliar foods.

Table 3. Preference of 'cecina' from different species (percentage; n = 145 consumers).

	1st Preferred	2nd Preferred	3rd Preferred
Horse	9.8	10.4	19.5
Cow	29.4	12.6	14.3
Sheep	2.8	4.4	4.5
Goat	4.9	4.4	6.8
Rabbit	-	0.7	1.5
Deer	26.6	31.9	23.3
Wild pig	12.6	28.1	15.0
Pig	8.4	5.2	12.0
Duck	5.6	2.2	3.0

3.2. Consumer Acceptability

The addition of different levels of linseed or the different duration of the feeding period did not modify ($p > 0.05$) the colour acceptability of 'cecina' (Table 4). This attribute was scored between 6.41–6.58 on a 9-point scale, which would be equivalent to answers like slightly to moderate. In the current data, no variation in colour would be expected, because the process of elaboration was identical for all treatments. Even the modification of some procedures, such as tumbling treatments [5] reported no significant differences in instrumental colour variables such as lightness, redness or yellowness. Consequently, it was expected that consumers did not report differences in colour acceptability. Differences on 'cecina' colour, such as those reported by [12] in beef, were associated with the different durations of drying.

Acceptability of fat quantity on 'cecina' samples from 15% of supplementation presented lower values than 5% or 10% ($p < 0.001$), which correspond to a Body Condition Score (BCS) of 3.33, probably considered the quantity of fat excessive. However, with regard to feeding duration ($p < 0.01$), samples from 30 days presented lower values than those from 50 days, having longer fattening periods (70 days) and intermediate scores. Products from the 30 days (BCS of 2.42) were probably considered as too lean.

On the sensorial attributes evaluated (odour, flavour and overall acceptability) linseed level was a significant factor ($p < 0.01$), with the 15% level presenting the lowest acceptabilities, but feeding duration did not modify ($p > 0.05$) those variables. However, there were significant interactions ($p < 0.01$) between both factors (linseed level and feeding duration).

Odour is an important attribute in dry cured products. As [5] reported, the most numerous compounds in the dry-cured lamb legs are straight-chain aldehydes followed by alcohols, ketones and benzene compounds. These compounds have been described in volatile profiles of dry-cured ruminant or pork meat products [1,21]. Variations in the ripening time and tumbling modified the presence of several compounds, such as 3-methylbutanal from leucine and phenylacetaldehyde from phenylanine. These compounds are considered to be major contributors of dry-cured ham flavour. The highest value for odour acceptability was associated with 'cecina' with 10% linseed supplementation and 30 days of fattening, with 6.29 points on a 9-point scale. This value did not differ statistically from those scored at any feeding period of the lowest level of supplementation 5%, or from those with 70 days of fattening but with 10% or 15% supplementation. The lowest acceptabilities were reported for 15% with 30 or 50 days of fattening.

In dry-cured products such as ham, taste is considered to be a key attribute influencing consumer overall liking, and colour, flavour and adequate saltiness are considered basic quality signals by consumers [4]. In the current study, differences in colour or saltiness did not exist, due to the identical processing. Flavour and overall acceptability presented the same consumer evaluation behaviour, and in meat it is also common to find a high correlation between the sensory attribute flavour and the overall acceptability of the evaluated product [22], which explains the equivalent behaviour of both variables. The preferred combination of treatments that present the highest punctuation (6.11 and 6.16, respectively, for flavour and overall acceptability) was 'cecina' from the lowest linseed supplementation and the lowest feeding duration (5% and 30 days). However, those values did not statistically differ from those of any of the combinations of feeding duration with 5 or 10% linseed, or 15% at 70 days. The lowest score came from 'cecina' with 15% and 30 days of fattening.

Volatile Free Fatty Acids (FFA) are formed by lipolysis, and FFA alkyl esters seem to be formed by microbial esterification during ripening [1,5]. Given that those compounds are associated with ripened flavour in cured meat products and increase through the ripening [5], they contribute directly to taste and indirectly to odour and flavour by means of aroma compounds [12].

Differences in odour and flavour associated with linseed level are probably linked to the modifications in the fatty acid profile produced by linseed supplementation [7], occurring as an increment in the percentage of total PUFA and n-3 PUFA. The levels of lipid oxidation-derived compounds are relevant for the flavour of dry-cured meat products [5].

In a general way, it can be observed that the overall acceptability of 'cecina' presented similar values (slighter lower) than those reported by [2] for Spanish consumers (6.00–6.83 points). Therefore, 'cecina' from cull ewes would presumably present an adequate acceptability and good commercial reception in Spain.

According to data from the current and the previous study [2], 'cecina' with lean or medium fat levels is preferred by consumers, so it is not necessary to have long feeding durations for finishing cull ewes (30 days would be enough). Also, low levels of linseed (5%) would be enough if the final product destination were the elaboration of 'cecina', or even up to 10%, if the aim were the consumption of fresh meat [7].

3.3. Cluster Analyses

Consumers present different preferences, which leads to the existence of several market niches. Three clusters (groups) of consumers with different preferences related to the overall acceptability of 'cecina' were identified (Table 5).

For the first cluster identified, only linseed level had an influence ($p < 0.001$) on the overall acceptability of 'cecina'. This cluster represented 14.7% of consumers in the study, and the main characteristics of this group were the rejection of the product, presenting low acceptability scores ranging between (2.47–3.30 points), especially 'cecina' from the highest level of linseed supplementation (15%), which presented the lowest score. In this group of people, the percentage of male consumers was 45.5% and the age of participants was almost equivalent for each of the 3 age groups, with consumers younger than 30 being slightly more abundant, 36.4% vs. 31.8% for the other two groups. Related to answers of the questionnaire, in this group, 90.9% of consumers eat 'cecina' very sporadically. Also, 90.9% of the group would pay less than 15 €/kg (which was the prize of reference for ham) for 'cecina', and none of the participants would also have preferred the 'cecina' of sheep in one of the top 3 species preferences, although only 13.6% thought that this type of 'cecina' existed in the market.

Cluster 2 included the most participants (58.0% of the sample). For this group the linseed level was a significant factor ($p < 0.05$) in acceptability, with 10% supplementation being preferred with respect to the other supplementation levels. However, although feeding duration did not present statistical differences, there was a significant interaction between both factors ($p < 0.001$). The overall acceptability score for all treatments in this group was good (from 6.71 to 7.22), higher than the average overall score. The highest scores were for the 10% supplementation level group with 30 or 70 days of fattening, and 15% with 50 days of fattening. The worst score treatment was for the 5% supplementation group with 50 days of fattening; but even then, that score was 6.71 points. With regard to socio-demographic characteristics, the cluster was formed of 47.1% men, with the most abundant age range being between 31 and 50 years (43.7%), and the least predominant consumers being those younger than 30 years (20.7%). According to the survey, 20.7% of this group consumed 'cecina' at least one at month, which showed in that this cluster was more familiar with the product than the other clusters, and explained the higher scores for 'cecina' than cluster 1. 3.53.6% of the cluster would pay a price equivalent to those given as reference for ham (15 €/kg: 20.7% of them) for 'cecina', or higher, up to 19.5 €/kg. 39% of these consumers think that 'cecina' from sheep exists on the market and 16.6% of answers indicate that this 'cecina' would be one of the top 3 preferred.

The last group identified included 27.3% of participants. Similar to cluster 2, linseed level was a significant factor ($p < 0.05$); however, the least preferred level of supplementation was 15%, with the acceptabilities for supplementation at 5 or 10% being similar. Also, there was a significant interaction between linseed supplementation level and fattening duration ($p < 0.001$). 'Cecina' from treatment with 15% linseed for 30 or 50 days, as well as, 'cecina' from 10% and 70 days of finishing, had the lowest acceptabilities of the different clusters, with scores lower than 5 points. Averages for acceptability in this cluster varied from 4.20 to 6.00 points. Cluster 3 was composed of a minority of men (43.9%). Consumers between 31–50 years old accounted for 63.4%, and older than 51 represented 17.1% of the cluster, and 19.5% were younger than 30 years. With regard to the frequency of 'cecina' consumption, in spite of the majority of people having 'cecina' very sporadically (85.4%), there were 4.9% that consumed the product once a month. 34.2% of consumers would pay more than 15 €/kg for 'cecina' and 53.6% would pay less than the price of reference (<15 €/kg). 7.4% of participants marked 'cecina' from sheep as one of the three preferred species of 'cecina' to consume.

Table 4. Acceptability of sensory attributes of 'cecina' from cull ewes fed with different linseed levels and fattening periods (n = 150 consumers).

	LSL			FD			5% LSL			10% LSL			15% LSL						
	5%	10%	15%	30 d	50 d	70 d	30 d	50 d	70 d	30 d	50 d	70 d	30 d	50 d	70 d	SEM	LSL	FD	LSL × FD
Colour	6.41	6.58	6.46	6.48	6.46	6.51	-	-	-	-	-	-	-	-	-	0.045	0.109	0.859	0.475
Fatness	6.58 ^{b}	6.65 ^{b}	6.36 ^{a}	6.44 ^{a}	6.65 ^{b}	6.50 ^{a,b}	-	-	-	-	-	-	-	-	-	0.043	0.001	0.034	0.307
Odour	-	-	-	-	-	-	6.06 ^{bc}	6.20 ^{b,c}	6.08 ^{bc}	6.29 ^{c}	5.86 ^{a,b}	5.95 ^{a,b,c}	5.66 ^{a}	5.66 ^{a}	6.04 ^{b,c}	0.053	0.001	0.384	0.004
Flavour	-	-	-	-	-	-	6.11 ^{c}	5.75 ^{b,c}	5.83 ^{bc}	5.95 ^{b,c}	5.90 ^{b,c}	5.85 ^{b,c}	5.29 ^{a}	5.63 ^{b}	5.90 ^{b,c}	0.059	0.003	0.580	0.002
Overall acceptability	-	-	-	-	-	-	6.16 ^{c}	5.88 ^{b,c}	5.94 ^{bc}	6.11 ^{bc}	5.97 ^{b,c}	5.84 ^{b,c}	5.39 ^{a}	5.76 ^{b}	6.03 ^{b,c}	0.056	0.004	0.736	0.000

$^{a–c}$, Lowercase letters indicate statistical differences in the same row between treatments ($p \leq 0.05$). SEM: standard error of mean; LSL: effect of linseed supplementation level (p-value); FD: effect of feeding duration (p-value); LSL × FD: interaction between linseed supplementation level and feeding duration (p-value). Based on a 9-point scale: (1: dislike extremely; 9: like extremely).

Table 5. Overall acceptability scores of 'cecina' from cull ewes fed with different linseed levels and fattening periods in three clusters of consumers (n = 150).

		LSL			FD			5% LSL			10% LSL			15% LSL						
	%	5%	10%	15%	30 d	50 d	70 d	30 d	50 d	70 d	30 d	50 d	70 d	30 d	50 d	70 d	SEM	LSL	FD	LSL × FD
Cluster 1	14.7	3.30 ^{b}	2.94 ^{b}	2.47 ^{a}	2.98	2.83	2.89	-	-	-	-	-	-	-	-	-	0.095	0.000	0.725	0.661
Cluster 2	58.0				-	-	-	7.08 ^{b,c}	6.71 ^{a}	6.97 ^{a,b,c}	7.20 ^{c}	7.02 ^{a,b,c}	7.22 ^{c}	6.74 ^{a,b}	7.16 ^{c}	6.92 ^{a,b,c}	0.042	0.029	0.763	0.010
Cluster 3	27.3				-	-	-	5.66 ^{c}	5.54 ^{c}	5.22 ^{b,c}	5.37 ^{c}	5.46 ^{c}	4.51 ^{a,b}	4.20 ^{a}	4.54 ^{ab}	6.00 ^{c}	0.098	0.046	0.752	0.00

$^{a–c}$, Lowercase letters indicate statistical differences in the same row between treatments ($p \leq 0.05$). SEM.: standard error of mean; LSL: effect of linseed supplementation level (p-value); FD: effect of feeding duration (p-value); LSL × FD: interaction between linseed supplementation level and feeding duration (p-value). Based on a 9-point scale: (1: dislike extremely; 9: like extremely).

4. Conclusions

According to the findings from the current study, 'cecina' from cull ewes presented an adequate acceptability for local consumers. The duration of the finishing period scarcely affected 'cecina' attributes and its acceptability, with the effect of linseed supplementation level being more significant. Low levels of fat are usually preferred by consumers, therefore short finishing periods (30 days) are enough to improve and homogenize the carcasses of cull ewes, and supplementations with low levels of linseed (5% or 10%) would be recommended based on 'cecina' attributes and economical profitability. This product would increase the income from adult animals, which usually have a low market value. Consequently, production of 'cecina' would be a feasible strategy to improve the yields of animals and income of farmers, having a positive economic outcome for breeders.

Author Contributions: F.A.F.M. and M.d.M.C. conceived and designed the experiment; all authors were involved in the development of the experiment and sensory analyses; A.G. and M.d.M.C. analysed the data, A.G., C.S., J.L.O., M.d.M.C. prepared the manuscript. All authors were involved in manuscript revisions and have read and approved the manuscript.

Funding: This work was supported by Zaragoza University (UZ2011-CIE-229-378).

Acknowledgments: The authors also thank Pastores Cooperative Group, Esperanza Horcas and Antonio Oliván and Secadero Sierra Maestrazgo for its technical assistance.

Conflicts of Interest: The authors declare no conflict of interest.

References

1. Hierro, E.; de la Hoz, L.; Ordóñez, J.A. Headspace volatile compounds from salted and occasionally smoked dried meats (cecinas) as affected by animal species. *Food Chem.* **2004**, *85*, 649–657. [CrossRef]
2. Sañudo, C.; Monteiro, A.L.G.; Valero, M.V.; Fugita, C.A.; Monge, P.; Guerrero, A.; Campo, M.M. Cross-cultural study of dry-cured sheep meat acceptability by native and immigrant consumers in Spain. *J. Sens. Stud.* **2016**, *31*, 12–21. [CrossRef]
3. Bernués, A.; Riedel, J.L.; Asensio, M.A.; Blanco, M.; Sanz, A.; Revilla, R.; Casasús, I. An integrated approach to studying the role of grazing livestock systems in the conservation of rangelands in a protected natural park (Sierra de Guara, Spain). *Livest. Prod. Sci.* **2005**, *96*, 75–85. [CrossRef]
4. Andrade, J.C.; Nalério, E.S.; Giongo, C.; Barcellos, M.D.; Ares, G.; Deliza, R. Consumer sensory and hedonic perception of sheep meat coppa under blind and informed conditions. *Meat Sci.* **2018**, *137*, 201–210. [CrossRef] [PubMed]
5. Villalobos-Delgado, L.H.; Caro, I.; Blanco, C.; Morán, L.; Prieto, N.; Bodas, R.; Giráldez, F.J.; Mateo, J. Quality characteristics of a dry-cured lamb leg as affected by tumbling after dry-salting and processing time. *Meat Sci.* **2014**, *97*, 115–122. [CrossRef] [PubMed]
6. Bhatt, R.S.; Soren, N.M.; Sahoo, A.; Karim, S.A. Level and period of realimentation to assess improvement in body condition and carcass quality in cull ewes. *Trop. Anim. Health Prod.* **2013**, *45*, 167–176. [CrossRef] [PubMed]
7. Guerrero, A.; Sañudo, C.; Campo, M.M.; Olleta, J.L.; Muela, E.; Macedo, R.M.G.; Macedo, F.A.F. Effect of linseed supplementation level and feeding duration on performance, carcass and meat quality of cull ewes. *Small Rum. Res.* **2018**, in press.
8. Manso, T.; Gallardo, B.; Guerra-Rivas, C. Modifying milk and meat fat quality through feed changes. *Small Rumin. Res.* **2016**, *142*, 31–37. [CrossRef]
9. Ragni, M.; Toteda, F.; Tufarelli, V.; Laudadio, V.; Facciolongo, A.; Dipalo, F.; Vicenti, A. Feeding of extruded flaxseed (*Linum usitatissimum* L.) and pasture in podolica young bulls: Effects on growth traits, meat quality and fatty acid composition. *Pakistan J. Zool.* **2014**, *46*, 1101–1109.
10. Tuorila, H. From sensory evaluation to sensory and consumer research of food: An autobiographical perspective. *Food Qual. Prefer.* **2015**, *40*, 255–262. [CrossRef]
11. Sañudo, C. *Atlas Mundial de Etnología Zootécnica*; Servet, Z., Ed.; Servet: Zaragoza, Spain, 2011; p. 399, ISBN 978-84-92569-60-1.

12. Molinero, C.; Martinez, B.; Rubio, B.; Rovira, J.; Jaime, I. The effects of extended curing on the microbiological, physicochemical and sensorial characteristics of Cecina de Leon. *Meat Sci.* **2008**, *80*, 370–379. [CrossRef] [PubMed]
13. Boutrolle, I.; Arranz, D.; Rogeaux, M.; Adelarue, J. Comparing central location test and home use test results: Application of a new criterion. *Food Qual. Prefer.* **2005**, *16*, 704–713. [CrossRef]
14. Instituto Nacional de Estadística. Available online: http://www.ine.es (accessed on 7 May 2018).
15. Furnols, M.; Realini, C.E.; Guerrero, L.; Oliver, M.A.; Sañudo, C.; Campo, M.M.; Nute, G.R.; Cañeque, V.; Álvarez, I.; San Julián, R.; et al. Acceptability of lamb fed on pasture, concentrate or combinations of both systems by European consumers. *Meat Sci.* **2009**, *81*, 196–202. [CrossRef] [PubMed]
16. Resano, H.; Sanjuán, A.I.; Albisu, L.M. Consumers' acceptability of cured ham in Spain and the influence of information. *Food Qual. Prefer.* **2007**, *18*, 1064–1076. [CrossRef]
17. MAPAMA. Available online: www.mapama.gob.es/es/alimentacion/temas/consumo-y-comercializaciondistribucionalimentaria/informe_del_consumo_de_alimentos_en_espana_2016_webvf_tcm30-419505.pdf (accessed on 7 May 2018).
18. Resano, H.; Pérez-Cueto, F.J.A.; Sanjuán, A.I.; Barcellos, M.D.; Grunert, K.G.; Verbeke, W. Consumer satisfaction with dry-cured ham in five European countries. *Meat Sci.* **2011**, *87*, 336–343. [CrossRef] [PubMed]
19. Andrade, J.C.; Nalério, E.S.; Giongo, C.; Barcellos, M.D.; Ares, G.; Deliza, R. Consumer perception of dry-cured sheep meat products: Influence of process parameters under different evoked contexts. *Meat Sci.* **2017**, *130*, 30–37. [CrossRef] [PubMed]
20. Tuorila, H.; Meiselman, H.L.; Bell, R.; Cardello, A.V.; Johnson, W. Role of Sensory and Cognitive Information in the Enhancement of Certainty and Linking for Novel and Familiar Foods. *Appetite* **1994**, *23*, 231–246. [CrossRef] [PubMed]
21. Narváez-Rivas, M.; Gallardo, E.; León-Camacho, M. Analysis of volatile compounds from Iberian hams: A review. *Grasas Aceites* **2012**, *63*, 432–454. [CrossRef]
22. Guerrero, A.; Campo, M.M.; Cilla, I.; Olleta, J.L.; Alcalde, M.J.; Horcada, A.; Sañudo, C. A comparison of laboratory-based and home-based tests of consumer preferences using goat and lamb meat. *J. Sens. Stud.* **2014**, *29*, 201–210. [CrossRef]

Article

Consumer Perception of the Quality of Lamb and Lamb Confit

Guillermo Ripoll *, Margalida Joy and Begoña Panea

Centro de Investigación y Tecnología Agroalimentaria de Aragón (CITA), Instituto Agroalimentario de Aragón—IA2 (CITA-Universidad de Zaragoza), Avda. Montañana 930, 50059 Zaragoza, Spain; mjoy@aragon.es (M.J.); bpanea@aragon.es (B.P.)

* Correspondence: gripoll@aragon.es; Tel.: +34-976-716-452

Received: 25 April 2018; Accepted: 21 May 2018; Published: 22 May 2018

Abstract: The patterns of food consumption in general and those of meat, in particular, are constantly changing. These changes are due not only to socioeconomic and cultural trends that affect the whole society but also to the specific lifestyles of consumer groups. Due to the importance of consumer lifestyle, the objectives of this study were (i) to identify the profiles of lamb meat consumers according to their orientation toward convenience, as defined by their eating and cooking habits; (ii) to characterize these profiles according to their socioeconomic characteristics and their preferences regarding the intrinsic and extrinsic quality signals of lamb meat; and (iii) to analyze the willingness to pay for lamb confit. In this study, four types of consumers have been differentiated according to their lifestyles related to lamb consumption. These groups, due to their characteristics, could be called "Gourmet", "Disinterested", "Conservative", and "Basic". The Gourmet group has characteristics that make it especially interesting to market a product such as lamb confit. However, this group is unaware of this product. Therefore, a possible strategy to expand the commercialization of light lamb and the confit product would be guided marketing to this niche market.

Keywords: cluster; intrinsic; extrinsic; oil; meat confit; lamb

---

1. Introduction

The patterns of food consumption, in general, and those of meat, in particular, are constantly changing. These changes are due not only to socioeconomic and cultural trends that affect the whole society but also to the specific lifestyles of consumer groups, which are increasingly diversified. Lifestyle defines the activities of people in terms of how they spend their time, their interests, their opinions, and their views of themselves and the world around them [1]. This makes the consumer respond differently to everyday stimuli such as consumption. In the field of food marketing research, Brunsø and Grunert [2] established as a methodological framework the food-related lifestyle (FRL) as a mediator between consumer values and their behavior. In this framework, an FRL is composed of five elements or interrelated aspects: how to make purchases, quality aspects for the evaluation of food products, cooking methods, consumption situations, and the reasons for purchase [3]. Bernués, et al. [4] used the methodology of FRL to perform a consumer segmentation regarding fresh lamb meat as convenience food. These convenience foods are increasingly demanded due to the greater incorporation of women into the labor market, the proliferation of small, single-parent or single-person families, individualistic and impulsive consumer behavior, and lack of interest or skill in the kitchen [5]. The four groups of consumers detected were those with a traditional profile, middle-aged people who like traditional foods and exhibit great concern for the intrinsic quality attributes of lamb; people little involved with food but who are aware of its importance and its relationship with health and who are concerned that lamb is easy to cook; the adventurers, with a very open attitude regarding innovation in their diet and a lot of interest in both intrinsic and extrinsic quality of lamb; and finally, the carefree, who are not interested in anything related to food and who are not interested in any particular characteristic of lamb [5]. However, these authors did not conduct any study regarding lamb products with a clear orientation toward convenience purchases, as is the case of light lamb confit and meat that is ready to be consumed after brief and minimal handling. Traditionally, the process of confit is applied to fruits and other products that are cooked in a syrup at low temperature over long cooking times. In recent years, the concept of confit has been extended to cooking in oil at low temperature for long periods, unlike frying, which involves cooking at high oil temperatures for a very short time. An internationally known example is duck confit, in which duck meat confits in its own fat. Light lamb confit is prepared in olive oil at low temperature and is presented as canned in oil. With the production of light lamb confit, which is currently not on the market, the breeders of light lamb would diversify the production and overcome seasonality in the sale of lamb. The diversification of lamb products is interesting due to the constant decrease of fresh lamb in Spain, especially in Aragon [6]. Aragón is an utmost region in the light lamb production and that decrease of meat from light lam consumption is quite harmful. In addition, Ojinegra de Teruel breed is one of the important breeds raised in Aragon. This breed had an early deposition of fat [7], which made it difficult to commercialize it as fresh meat.

Light lamb confit has a flavor somewhat different from lamb cooked by other means and is a product that must simply be heated before serving. As the characteristic flavor of lamb and the difficulty of cooking it are two of the factors that have caused the consumption of lamb to decrease in Spain [8], it is possible that this presentation of light lamb confit, canned with oil, will increase consumption. Currently, the convenience of cooking and consumption is increasingly perceived as a key factor in the marketing of any meat product and, especially for light lamb meat, is closely linked to the manners in which it is cooked and consumed. Therefore, it can be deduced that the future consumption of light lamb meat, or its subsequent replacement by other meats, will depend to a great extent on these two elements: the manners in which it is cooked and consumed. Therefore, the present study has focused on these two aspects to establish consumer orientation toward convenience. This approach is consistent with the "supply, perception and demand for quality" model of food established by [3,4] for the segmentation of consumers.

The objectives of this study were (i) to identify the profiles of light lamb meat consumers according to their orientation toward convenience, as defined by their eating and cooking habits; (ii) to characterize these profiles according to their socioeconomic characteristics and their preferences regarding the intrinsic and extrinsic quality cues of light lamb meat; and (iii) to analyze the willingness to pay for light lamb confit.

2. Materials and Methods

An online survey was conducted using forms from Google, Inc. (Menlo Park, CA, USA) during the months of May and June 2014. The geographical scope of the survey was restricted to Spain, ruling out the responses of consumers from other countries. This study was conducted according to the Declaration of Helsinki for studies on human subjects.

The survey consisted of four blocks: (A) socio-demographic variables (sex, age, level of income per capita, level of schooling, etc.); (B) variables related to lifestyle, specifically regarding habits related to eating and cooking; (C) the importance of extrinsic quality attributes of lamb meat; and (D) the importance of the intrinsic quality attributes of lamb meat at the time of purchase, scored according to a 4-point Likert-type scale ((1) not at all important; (2) not very important; (3) fairly important; (4) very important). Blocks B, C and D are shown in Table 1. For the statements in (B) and (C), the respondent had to express his/her degree of agreement or disagreement on a 4-point Likert-type scale ((1) strongly disagree; (2) disagree; (3) agree; (4) strongly agree).

Table 1. Questionnaire about consumer habits when cooking and eating and the importance of extrinsic and intrinsic attributes of lamb meat.

(B) Habits of the consumer at the time of cooking and eating [1]
I like to cook
I like foreign food
At home, we prefer informal dinners
I like going to restaurants with friends and family
At home, everyone cooks
I spend a lot of time cooking
I like to eat
I like changes in food
Planning food is important for family nutrition
Traditional recipes are best
(C) Importance of extrinsic quality attributes of lamb [1]
Lamb meat with a mark of quality is better
Lamb is easy to cook
The best lamb is grass-fed
The Ojinegra de Teruel breed is better than others
The price of lamb is very important
The lamb from Aragon is better than others
Organic lamb is better than others
The best lamb is fed cereals
(D) Importance of the intrinsic quality attributes of lamb meat at the time of purchase [2]
Appearance of freshness
Lamb category
Light lamb category
Light colored meat
Age
Breed
Low fat
(E) Other issues related to lifestyles
Are you vegetarian? [3]
Do you have any food restrictions for your religion? [3]
Would you like to lose weight? [3]
Is your cholesterol level high? [3]
Do you live in a city? [3]
Have you heard about the Ojinegra de Teruel breed? [3]
Have you heard about lamb confit? [3]
Have you ever eaten lamb confit? [3]
If you have eaten lamb confit, where did you eat it? [4]
How much would you be willing to pay for a can for four people of lamb confit preserved in ... [5]
- Extra virgin olive oil
- Olive oil
- Sunflower oil

[1] 4-point Likert scale, Strongly disagree, Disagree, Agree and Strongly agree. [2] 4-point Likert-type scale, Not important, A little important, Quite important, Very important. [3] Yes/No. [4] I cooked it myself, In a restaurant, I bought it made, Other answer. [5] Less than 10 euros, between 10 and 15 euros, More than 15 euros.

In addition to the questions related to the place of residence and health ("Would you like to lose weight?" and "Is your cholesterol level high?"), respondents were asked about the price they would be willing to pay for a can of light lamb confit for four people that was preserved in three different types of oil (extra virgin olive, virgin olive or sunflower oil). The options were as follows: less than 10 euros, between 10 and 15 euros and more than 15 euros. In the absence of such a product on the market, these proposed prices are based on confit meats of other types of lamb and other meats in general. Due to not having a closed interval as a price reference, the lower and upper ends were left open.

Except for the variable Age, which was continuous, all the questions in the survey were closed and of an ordinal type (blocks B, C and D) or nominal.

Once the survey was available online, the access link was disseminated via email to both individuals and institutions and groups (housewives, consumers, cultural associations, etc.). Social networks such as Facebook, Twitter and personal blogs were also used. At the end of the two-month duration of the survey period, 659 surveys had been collected, of which 200 surveys were complete and corresponded to Spanish consumers.

The study of the consumer sample was performed by means of relative frequencies. The analysis of the variables was performed using the χ^2 test, taking a probability less than 0.10 as significant. When one of the cells had a frequency less than five, which makes the use of the χ^2 statistic unadvisable, the likelihood ratio statistic was used at the same probability level. To interpret the pattern of association between the variables studied, the corrected standardized residual between the observed and expected cases within each cell greater than $|1.96|$ was considered. The corresponding percentage associated with these residuals is specified in bold in the tables.

To group consumers into homogeneous groups (clusters or conglomerates), hierarchical cluster analysis was performed using the Ward method. In this analysis, the variables of the B blocks were included, following the theoretical framework of FRL proposed by Brunsø and Grunert [2] and developed by Grunert [9]. The number of clusters or groups of consumers by affinity of their chosen responses was a compromise solution using Ward's distance from the dendrogram that would maximize the distance between one division of the dendrogram and the next, so as to not obtain a number of clusters too great to be discussed. Subsequently, the relationships between the different groups were analyzed using the test χ^2 test or the likelihood ratio under the conditions discussed above.

3. Results

3.1. Characterization of the Sample

The data for the general sample are presented in Tables 2–6. The sample (Table 2) had 5% more women, and there was a bias in the level of studies because 50% of the respondents had university studies. Regarding the level of income per capita, less than 25% earned less than €1000/month, and the rest of the sample had income distributed similarly among the three upper strata. Although the survey was initially disseminated throughout Spain and responses were even received from other countries, by not requesting the region of origin of the applicant, we cannot confirm a geographical bias within Spain.

We found that 1.1% of respondents defined themselves as vegetarians (Table 3) and 2.7% held religious beliefs with certain types of food restrictions. Approximately one-third of the sample knew of the Ojinegra de Teruel breed and lamb confit. A large percentage (70.9%) did not know whether they had ever eaten confit, and 12.5% had never tasted it. Only 16.6% answered that they knew of it, of which 72.7% knew it because they had eaten it in a restaurant (Table 3).

Table 2. Socio-demographic characteristics of the general sample and by consumer group.

	CL1	CL2	CL3	CL4	Total	χ^2	p
Respondents/Percentage of Respondents	**82/41%**	**4/2%**	**66/33%**	**48/24%**	**200**		
Sex						6.19	t
Man	41.8	**0.0**	49.2	**59.1**	47.6		
Woman	58.2	**100.0**	50.8	**40.9**	52.4		
Level of schooling						50.87	***
None	0.0	**33.0**	1.6	0.0	1.1		
Primary	0.0	0.0	0.0	2.3	0.5		
Secondary	1.3	**33.3**	3.3	4.5	3.2		
High School	19.0	0.0	21.3	20.5	19.8		
Vocational training	19.0	0.0	26.2	34.1	24.6		
University students	**60.8**	33.3	47.5	38.6	50.8		
Is vegetarian	0.0	**33.3**	1.6	0.0	1.1	31.03	***
Income per capita						17.14	ns
<600 €/month	15.6	**33.3**	5.4	12.2	11.9		
600—1000 €/month	9.1	**33.3**	12.5	22.0	13.6		
1000—1500 €/month	27.3	**33.3**	23.2	12.2	22.6		
1500—2000 €/month	24.7	**0.0**	23.2	31.7	25.4		
>2000 €/month	23.4	**0.0**	35.7	22.6	26.6		
Religious food restrictions	2.5	**33.3**	1.6	2.3	2.7	11.12	*

ns, $p > 0.10$; t, $p < 0.10$; *, $p < 0.05$; ***, $p < 0.001$. Cells in bold had the corrected standardized residual between the observed and expected cases within each cell greater than |1.96|.

Table 3. Other questions related to lifestyle.

	CL1	CL2	CL3	CL4	Total	χ^2	p
Respondents/Percentage of Respondents	**82/41%**	**4/2%**	**66/33%**	**48/24%**	**200**		
Are you Vegetarian? [1]	0.0	**33.3**	1.6	0.0	1.1	8.11	*
Do you have any food restrictions for your religion? [1]	2.5	33.3	1.6	2.3	2.7	3.86	ns
Would you like to lose weight? [1]	67.1	66.7	76.7	65.9	69.8	2.00	ns
Is your cholesterol level high? [1]	21.5	66.7	26.2	22.7	24.1	2.91	ns
Do you live in a city? [1]	81.0	100.0	82.0	77.3	79.4	1.67	ns
Have you heard of the Ojinegra de Teruel breed?	41.8	66.7	21.7	31.8	34.2	7.83	t
Have you heard of lamb confit? [1]	27.8	33.3	24.6	31.8	29.1	0.71	ns
Have you ever eaten lamb confit? [1,2]	54.5	100.0	46.7	71.4	16.6	3.06	ns
Where did you eat it? [3]							
I cooked it myself	30.0	0.0	0.0	0.0	16.0	10.14	ns
In a restaurant	66.7	100.0	85.7	90.0	72.7		
I bought it made	0.0	0.0	14.3	10.0	6.1		
Other answers	13.3	0.0	0.0	0.0	15.2		

[1] Percentage of affirmative responses. [2] The percentage of Do not know/No answer was 70.9%. [3] Percentage of respondents who claimed to have eaten lamb confit. Cells in bold had the corrected standardized residual between the observed and expected cases within each cell greater than |1.96|. ns, $p > 0.10$; *, $p < 0.05$.

Table 4 presents the degree of agreement with statements that define consumer lifestyle. Most respondents like to eat (96.8%), eat foreign food (72.8%), and prefer informal dinners (70.5%) and variety in food (74.4%), showing great interest in food and an open attitude to different foods and variety. There is a great agreement (greater than 93%) in terms of both the preference to eat at restaurants with friends and family and the importance of dietary planning for family nutrition.

Table 4. Habits of the consumer at the time of cooking and eating for the general sample and by groups of consumers.

	CL1	CL2	CL3	CL4	Total	χ^2	p
Respondents/Percentage of Respondents	**82/41%**	**4/2%**	**66/33%**	**48/24%**	**200**		
I like to cook						118.81	ns
Strongly disagree	0.0	66.7	14.8	2.3	6.4		
Disagree	1.3	33.3	47.5	0.0	16.6		
Agree	49.4	0.0	37.7	61.4	47.6		
Strongly agree	49.4	0.0	0.0	36.4	29.4		
I like foreign food						43.95	***
Strongly disagree	0.0	**33.3**	6.6	6.8	4.0		
Disagree	**5.1**	66.7	**37.7**	31.8	23.0		
Agree	**65.8**	0.0	**42.6**	50.0	53.5		
Strongly agree	**29.1**	0.0	13.1	11.4	19.3		
At home, we prefer informal dinners						49.45	***
Strongly disagree	8.9	**100.0**	**1.6**	9.1	8.0		
Disagree	20.3	0.0	13.1	**36.4**	21.4		
Agree	54.4	0.0	62.3	47.7	54.5		
Strongly agree	1.0	0.0	23.0	6.8	16.0		
I like going to restaurants with friends and family						132.77	***
Strongly disagree	0.0	**66.7**	0.0	0.0	1.1		
Disagree	3.8	**33.3**	4.9	6.8	5.3		
Agree	48.1	0.0	57.4	43.2	49.2		
Strongly agree	48.1	0.0	37.7	50.0	44.4		
At home, everyone cooks						48.36	***
Strongly disagree	**8.9**	**33.3**	32.8	**6.8**	16.6		
Disagree	**20.3**	**66.7**	42.6	43.2	33.7		
Agree	43.0	**0.0**	**19.7**	45.5	35.3		
Strongly agree	27.8	**0.0**	**4.9**	**4.5**	14.4		
I spend a lot of time cooking						72.17	***
Strongly disagree	**2.5**	0.0	**29.5**	1.7	11.8		
Disagree	**20.3**	0.0	**59.0**	38.6	36.9		
Agree	**55.7**	66.7	**8.2**	47.7	38.5		
Strongly agree	**21.5**	33.3	**3.3**	9.1	12.8		
I like to eat						143.40	***
Strongly disagree	1.3	**100.0**	0.0	2.3	2.7		
Disagree	0.0	0.0	1.6	0.0	0.5		
Agree	**26.6**	0.0	**72.1**	47.7	46.0		
Strongly agree	**72.2**	0.0	**26.2**	50.0	50.8		
I like changes in food						73.34	***
Strongly disagree	**0.0**	0.0	6.6	**11.4**	4.8		
Disagree	**3.8**	0.0	19.7	**54.5**	20.9		
Agree	43.0	33.3	54.1	31.8	43.9		
Strongly agree	**53.2**	66.7	19.7	**2.3**	30.5		
Planning meals is important for family nutrition						62.06	***
Strongly disagree	2.5	**66.7**	1.6	0.0	2.7		
Disagree	1.3	**33.3**	3.3	2.3	2.7		
Agree	39.2	0.0	39.3	34.1	37.4		
Strongly agree	57.0	0.0	55.7	63.6	57.2		
Traditional recipes are best						77.19	***
Strongly disagree	6.3	**66.7**	9.8	**0.0**	7.0		
Disagree	**51.9**	33.3	**54.1**	**11.4**	42.8		
Agree	41.8	0.0	**27.9**	47.7	38.0		
Strongly agree	**0.0**	0.0	8.2	**40.9**	12.3		

ns, $p > 0.10$; ***, $p < 0.001$. Cells in bold had the corrected standardized residual between the observed and expected cases within each cell greater than $|1.96|$.

Regarding extrinsic quality attributes (Table 5), the respondents agreed or strongly agreed that lamb meat with some mark of quality is better (78.4%). More than half thought that the lamb of Aragon (51.9%) or organic lamb (57.5%) is better than others, whereas only 32.1% agreed that the Ojinegra de Teruel breed is better than others; 70.4% believe that grass-fed lamb is better, whereas 29.9% think that cereal-fed lamb is best. The price of lamb is very important for 80.1% of respondents, and 68.7% think that lamb is easy to cook.

Table 5. Importance of the extrinsic attributes of lamb quality.

	CL1	CL2	CL3	CL4	Total	χ^2	p
Respondents/Percentage of Respondents	**82/41%**	**4/2%**	**66/33%**	**48/24%**	**200**		
Lamb meat with a mark of quality is better						15.88	t
Strongly disagree	3.8	33.3	5.0	2.3	4.3		
Disagree	25.3	0.0	11.7	11.6	17.3		
Agree	54.4	66.7	70.0	60.5	61.1		
Strongly agree	16.5	0.0	13.3	25.6	17.3		
Lamb is easy to cook						48.19	*
Strongly disagree	**1.3**	**100**	8.3	2.3	4.9		
Disagree	29.1	0	28.3	20.5	26.5		
Agree	55.7	0	50	50	51.9		
Strongly agree	13.9	0	13.3	**27.3**	16.8		
The best lamb is grass-fed						10.58	ns
Strongly disagree	7.6	33.3	5.0	2.3	5.9		
Disagree	20.3	66.7	23.3	27.3	23.7		
Agree	48.1	0.0	48.3	50.0	47.8		
Strongly agree	24.1	0.0	23.3	20.5	22.6		
The Ojinegra de Teruel breed is better than others						13.85	ns
Strongly disagree	8.9	33.3	10.9	9.8	10.1		
Disagree	65.8	33.3	56.4	46.3	57.9		
Agree	22.8	0.0	29.1	34.1	27.0		
Strongly agree	2.5	33.3	3.6	9.8	5.1		
The price of lamb is very important						25.20	**
Strongly disagree	1.3	**33.3**	1.6	2.3	2.2		
Disagree	17.7	33.3	24.6	7.0	17.7		
Agree	55.7	**0.0**	54.1	76.7	59.1		
Strongly agree	25.3	33.3	19.7	14.0	21.0		
The lamb from Aragon is better than others						14.550	ns
Strongly disagree	2.5	0.0	5.3	4.7	3.9		
Disagree	51.9	0.0	47.4	27.9	44.2		
Agree	25.3	50.0	36.8	37.2	32.0		
Strongly agree	20.3	50.0	10.5	30.2	19.9		
Organic lamb is better than others							
Strongly disagree	11.4	33.3	16.7	4.5	11.8	7.72	ns
Disagree	30.4	33.3	23.3	40.9	30.6		
Agree	43.0	33.3	46.7	40.9	43.5		
Strongly agree	15.2	0.0	13.3	13.6	14.0		
The best lamb is fed cereals						9.52	ns
Strongly disagree	11.4	33.3	6.8	11.6	10.3		
Disagree	60.8	0.0	67.8	51.2	59.8		
Agree	25.3	66.7	20.3	32.6	26.1		
Strongly agree	2.5	0.0	5.1	4.7	3.8		

ns, $p > 0.10$; t, $p < 0.10$; *, $p < 0.05$; **, $p < 0.01$. Cells in bold had the corrected standardized residual between the observed and expected cases within each cell greater than $|1.96|$.

The importance given to the intrinsic attributes of lamb at the time of purchase is listed in Table 6. The attributes valued as the most important were freshness (95.2%), age (84.5%), low fat content (70.4%)

and categorized as light lamb (67.2%). However, 58% of the respondents gave importance to the lamb category, 50.8% preferred lamb to have light-colored meat, and only 38.5% rated the breed as quite or very important at the moment of purchase.

Table 6. Importance of the intrinsic quality attributes of lamb meat at the time of purchase.

	CL1	CL2	CL3	CL4	Total	χ^2	p
Respondents/Percentage of Respondents	**82/41%**	**4/2%**	**66/33%**	**48/24%**	**200**		
Appearance of freshness						64.06	***
Nothing	0.0	**33.3**	0.0	0.0	0.5		
Little bit	2.5	0.0	6.6	4.5	4.3		
Quite	29.1	33.3	34.4	29.5	31.0		
A lot	68.4	33.3	59.0	65.9	64.2		
Lamb category						10.11	ns
Nothing	13.9	33.3	8.3	6.8	10.8		
Little bit	29.1	0.0	30.0	38.6	31.2		
Quite	34.2	66.7	46.7	29.5	37.6		
A lot	22.8	0.0	15.0	25.0	20.4		
Light lamb category						6.99	ns
Nothing	8.9	33.3	5.0	4.5	7.0		
Little bit	27.8	0.0	28.3	20.5	25.8		
Quite	41.8	33.3	48.3	50.0	45.7		
A lot	21.5	33.3	18.3	25.0	21.5		
Light colored meat						14.03	ns
Nothing	6.3	33.3	1.7	7.0	5.4		
Little bit	44.3	0.0	50.0	37.2	43.8		
Quite	38.0	33.3	41.7	34.9	38.4		
A lot	11.4	33.3	6.7	20.9	12.4		
Age						14.48	ns
Nothing	3.8	33.3	0.0	2.3	2.7		
Little bit	11.4	0.0	13.1	15.9	12.8		
Quite	48.1	33.3	47.5	38.6	45.5		
A lot	36.7	33.3	39.3	43.2	39.0		
Breed							
Nothing	11.4	33.3	**1.6**	6.8	7.5	18.82	*
Little bit	**62.0**	0.0	52.5	45.5	54.0		
Quite	**16.5**	33.3	36.1	38.6	28.3		
A lot	10.1	33.3	9.8	9.1	10.2		
Low fat						11.78	ns
Nothing	5.1	33.3	1.7	2.3	3.8		
Little bit	22.8	0.0	31.7	25.0	25.8		
Quite	51.9	33.3	45.0	56.8	50.5		
A lot	20.3	33.3	21.7	15.9	19.9		

ns, $p > 0.10$; *, $p < 0.05$; ***, $p < 0.001$. Cells in bold had the corrected standardized residual between the observed and expected cases within each cell greater than $|1.96|$.

3.2. *Types of Consumers*

Once the general sample was characterized, 4 types of consumers were identified based on their lifestyle related to their cooking and eating habits (Table 2). The first group (CL1) included 82 respondents (41%), group 3 (CL3) included 66 respondents (33%), and group 4 (CL4) accounted for 48 respondents (24%); the second group (CL2) was the minority, representing only 2% of the sample population including 4 respondents. The different groups of consumers did not differ in terms of level of income or place of residence ($p > 0.10$).

Socio-demographic characteristics and lifestyles related to cooking and eating are presented in Tables 2 and 3, respectively. CL1 was made up of both women and men, most of them with university education. This group was distinct because the percentage of respondents who knew of the Ojinegra de Teruel breed was higher than the general average. In terms of their lifestyle, this group is consumers who love cooking and eating, they live in homes in which everyone cooks, and they spend a lot of time cooking. In addition, they like foreign recipes and variety, and accordingly, they do not believe that lifelong recipes are better. Regarding the importance given to the extrinsic attributes of the quality of lamb meat, this group of consumers exhibited greater disagreement than the general sample regarding which lamb meat with a mark of quality is better, and few strongly disagreed with the ease of cooking lamb. Although this group knew of the Ojinegra de Teruel breed, it did not have a different perception from the other groups regarding this breed having better quality. This response was in accordance with the fact that breed, as an intrinsic attribute of quality, seems to be an unimportant factor.

CL2 is a minority group, but it has a lifestyle very different from the other groups of consumers; it is formed by women with secondary education or without formal education. One-third of the consumers in this group are vegetarians, and another third have food restrictions because of their religious beliefs. Consumers belonging to this group have a clear disinterest in everything related to cooking and food habits. They do not like foreign food, but they do not think that traditional food recipes are better than modern recipes. They also stand out because they do not like dinners, and they do not cook at home. Compared to the general sample, a significantly greater fraction of this group strongly disagreed that branded lamb is better. All the respondents in this group disagreed strongly with the statement that lamb is easy to cook and that the price of lamb is important. They also considered the appearance of freshness, the age of the lamb and the light color of the meat as significant characteristics.

The third group (CL3) does not like to cook or devote time to cooking, but almost everyone likes to eat; at home, not everyone cooks equally. They do not prefer foreign food because they think that lifelong recipes are best. This group declared mostly not knowing of the Ojinegra de Teruel breed. This group does not stand out from the general sample in terms of the importance assigned to quality attributes, both intrinsic and extrinsic.

CL4 had a higher proportion of men than the other groups. They like to cook but without making changes, and 88.6% agree or strongly agree that traditional recipes are best. This group of consumers also do not like informal dinners. A total of 27.3% of these consumers agreed that lamb meat is easy to cook. Additionally, those that think that price is very important predominate, and there are fewer that disagree with the lamb from Aragon being better than the rest. This group assigned high importance to light-colored meat, although they do not assign more importance to the category (lamb or light lamb) than the other groups or the sample.

3.3. Willingness to Pay

There was no relationship between the consumer group and the willingness to pay for lamb confit in any type of oil. Of the general sample, almost 32% would pay more than €15 for a can of lamb confit preserved in extra virgin olive oil, and 52% would pay between €10 and €15 (Figure 1). The remaining 15.6% would only pay less than €10. The level of studies, income, sex and all other socio-demographic questions did not have a significant relationship with the willingness to pay for lamb confit ($p > 0.05$) (data not shown). A significant relationship was found ($p = 0.0362$) between consumers with high cholesterol and their willingness to pay. A total of 64.6% of consumers willing to pay between €10 and €15 and 6.3% of those willing to pay less than €10 had high cholesterol. Consumers who would pay more than €15 would do so regardless of whether they had high cholesterol. Consumers who previously knew of lamb confit had a greater willingness (44.4%) to pay more than €15 for lamb confit ($p = 0.052$).

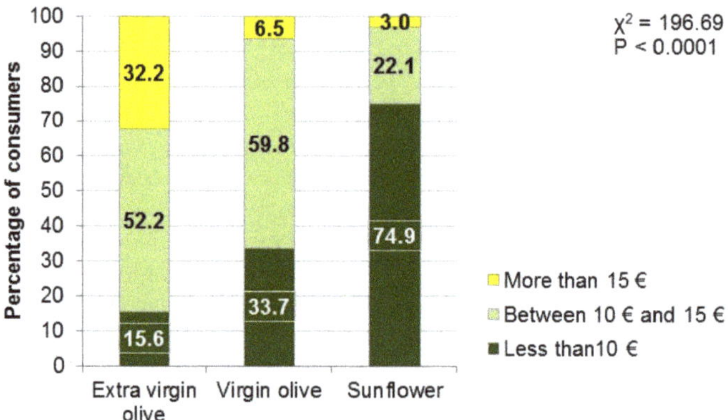

Figure 1. Willingness to pay depending on the type of oil used in the lamb confit.

Regarding questions related to consumer habits when cooking and eating, only the frequency with which the consumer eats outside the home ($p < 0.011$) and how important it is to plan meals for family nutrition ($p < 0.062$) affected willingness to pay. Consumers who eat away from home were less willing to pay for lamb confit. Regarding food planning, consumers who agree more about the importance of planning are more willing to pay. In fact, 67.2% of consumers who would pay more than €15 strongly agree about the importance of planning.

The importance assigned by the consumers to most extrinsic quality attributes did not determine the price that he or she would be willing to pay. It did affect the price that would be paid for the lamb to be of Aragonese origin ($p < 0.01$) and for organic lamb ($p < 0.01$); therefore, consumers who do not consider these two attributes important tend to be those who would pay less than €10 for the lamb confit. Consumers who value as important that the lamb they eat is categorized as light lamb ($p = 0.017$) are willing to pay more than €15, whereas consumers who do not give importance to this categorization are only willing to pay up to €10.

4. Discussion

According to Cotes [10], the multiple factors that affect the consumption decisions of an individual can be grouped based on his or her demographic characteristics. However, as seen in the results of this work, the behavior of the food consumer does not differ so much based on socio-demographic characteristics, but it does depend on their lifestyles [11]. The so-called psychographic characteristics include all the perceptions or beliefs of the individual, such as beliefs regarding the quality of a brand and propensity to value natural products. Thus, the opinion that each consumer has about the nutritional characteristics or the composition of a product or its safety, brand or price modify decisions at the time of purchase and even the degree of pleasure when it is consumed [12]. In short, consumers relate a group of products to a group of values through a system based on cognitive categories and actions that are embodied in a lifestyle [2].

4.1. Types of Consumers

It is possible to define numerous consumer groups; however, when these are framed regarding the lifestyle related to lamb consumption, the number of consumer segments in Spain is usually between three and five [4,13]. Although there are methodological differences in studies, the present work highlights the existence of four groups of differentiated consumers. These groups, due to their characteristics, could be called "Gourmet" (CL1), "Disinterested" (CL2), "Conservative" (CL3) and "Basic" (CL4). When Bredahl and Grunert [14] studied food lifestyles in Spain, they found five segments of consumers. The "Conservative" and "Uninvolved" segments of Bredahl and Grunert [14] are homologous to those found in this work and are called "Conservative" and "Disinterested", respectively. Consumers in the Conservative group like to cook but without making changes or innovations in the kitchen, which agrees with their opinion that traditional recipes are best. Another segment that the previous authors [4,14] called "Adventurers" would coincide with the gourmet group defined in this work, in that they are very fond of cooking, they use new recipes, and the whole family participates in culinary tasks. The segment called "Rational" by Bredahl and Grunert [14] would be comparable in many of its facets with our "Basic" segment. However, in the case of "Rational", the social role of food is particularly important, whereas for the "Basic" group, this facet is important but not more so than for other groups. Other studies have also found segments in other countries partially comparable with our results [15–17]. Of these recurrent segments in the literature, the one that represents the lowest percentage of the population is the Disinterested, which is formed mostly by women. Buckley, Cowan and McCarthy [15] found a group comparable to the disinterested, consumers who were primarily women who exhibited great individualism in their style of consumption. They do not plan meals or purchases, tend to consume snacks between meals and do not easily accept new products. The fact that women declare that they like exotic foods less agrees with the female gender being positively related to consumer ethnocentrism [18,19]. Thus, consumers with high ethnocentrism prefer to eat and buy products exclusively known and related to their culture. The low consumption of meat is generally a feminine phenomenon [20], and women have a greater number of food restrictions, especially regarding the consumption of red meat [21]. In fact, in a survey in Norway and Sweden, 72.5% of people who consumed little meat were women [22].

4.2. Importance of Intrinsic and Extrinsic Attributes

Regarding the importance of the attributes of fresh meat, among the most important according to the results of this study and others are low fat content, freshness and a defined brand or category [4,23,24], all of which are assessed visually. In contrast, breed and organic production are the least valued [4]. Low fat content is an important attribute for all consumers without differences between segments [4]. However, meat with a mark of quality, such as Denomination of Origin or Protected Geographical Indication, can be highly valued or valued very little depending on the consumer group [4], as observed in the present study. The region of origin of the product determines the importance given to some characteristics, even within a certain country. For example, within Spain, consumers in Castilla and León prefer ovine meat from lamb [25], whereas consumers in Aragon [4,26] prefer ovine meat categorized as light lamb. Although it has been seen that breed is not one of the important factors for the consumer, positive value has been given to meat that has been produced in the region where the consumer lives [3,9,26]. However, in our study, the lamb or light lamb category is considered important, but it is not considered that the Ternasco de Aragón PGI (light lamb) or the Ojinegra breed are better than other marks of quality or breeds. As seen in this study, price is a very important factor in general but also for some groups of consumers more than others. Thus, regular lamb consumers perceive higher quality due to marks of quality, diet or ecology. All this information associated with quality makes the price less relevant to the consumer than when there is less information [27].

4.3. Willingness to Pay

The process of perception of quality by consumers consists of two phases: the first is based on the perception of extrinsic and intrinsic attributes formed at the point of purchase, and the second is based on the experience formed during the preparation and consumption of the product [28,29]. This is when the expectations formed during the purchase are confirmed or rejected. These two stages are less important when a product is unknown (as may be the case of lamb confit because only a small part of the sample knew of it) or when the consumer does not know whether he or she has tasted it. In this case, the willingness to pay and even the decision to purchase the unknown product are probably based largely on the type of consumer. Thus, an "adventurous" consumer may pay more to try a product that he or she does not know, unlike more conservative or traditional consumers. However, in the scope of this work, there was no differential behavior between groups of consumers for willingness to pay for lamb confit. On the other hand, when the consumer is given information about the quality of the oil used in cooking, which is also a known product, perception of quality and price are positively related [27]. Moreover, although lamb confit is not a known product, olive oil is a well-known and traditional Mediterranean product, and prior knowledge of the product is important when buying food [30,31]. In addition, the appreciation of the different oils includes factors that take into account health [28]. The perception of a food as healthy (as is the case of virgin olive oil), natural, organic or respectful of the environment leads to a greater willingness to pay for the product [31,32], as can be clearly seen in the results of this study. However, when the consumer is not willing to pay a certain price, the conclusion is that the product does not have perceived net value [27], which would be the case of lamb confit preserved in sunflower oil, whose price is limited to a maximum of €15.

5. Conclusions

Market segmentation is a necessary requirement to ensure the creation of relationships between products and consumers. In this study, four types of consumers have been differentiated according to their lifestyle related to lamb consumption. One of these groups of consumers (gourmet) has characteristics that make it interesting to market a product such as lamb confit, considering they are unaware of this product. Therefore, a strategy to expand the commercialization of light lamb and the confit product would guide marketing to this niche market. To make the product known, a price strategy could be followed because although price is a clear indicator of perceived quality, when a consumer buys a product repeatedly, he or she gains experience, and the price has a lower weight as a key factor for the purchase. In fact, when the buyer is habitual, he or she values other factors, such as meat, with a mark of quality, even though it has a higher cost associated with it. As there is a direct relationship between the type of oil used in the processing of lamb confit and willingness to pay for the product, a range of products can be made based on this oil and the sale price. In the meat market, there is a range of opportunities that must be exploited with several strategies. In the case of lamb confit, it may be the promotion of the breed as an extrinsic attribute to transmit a higher-quality product.

Author Contributions: Conceptualization, G.R. and B.P.; Methodology, G.R. and B.P.; Formal Analysis, G.R.; Writing-Original Draft Preparation, G.R., M.J. and B.P.; Writing-Review & Editing, G.R., M.J. and B.P.; Funding Acquisition, M.J.

Funding: This research was funded by Investment Fund in Teruel (FIT) of the Government of Spain (333). B.P. and G.R. Authors are members of the MARCAME network, funded by CYTED (ref. 116RT0503).

Acknowledgments: The authors are grateful for the collaboration with Agroji and Cenro.

Conflicts of Interest: The authors declare no conflict of interest.

References

1. Plummer, J.T. Concept and application of life style segmentation. *J. Mark.* **1974**, *38*, 33–37. [CrossRef]
2. Brunsø, K.; Grunert, K.G. Development and testing of a cross-culturally valid instrument: Food-related life style. *Adv. Consum. Res.* **1995**, *22*, 475–480.
3. Bernués, A.; Olaizola, A.; Corcoran, K. Extrinsic attributes of red meat as indicators of quality in Europe: An application for market segmentation. *Food Qual. Prefer.* **2003**, *14*, 265–276. [CrossRef]
4. Bernués, A.; Ripoll, G.; Panea, B. Consumer segmentation based on convenience orientation and attitudes towards quality attributes of lamb meat. *Food Qual. Prefer.* **2012**, *26*, 211–220. [CrossRef]
5. De Boer, M.; McCarthy, M.; Cowan, C.; Ryan, I. The influence of lifestyle characteristics and beliefs about convenience food on the demand for convenience foods in the Irish market. *Food Qual. Prefer.* **2004**, *15*, 155–165. [CrossRef]
6. MERCASA. *Alimentación en España 2015. Producción, Industria, Distribución y Consumo*; MERCASA: Madrid, Spain, 2016; p. 552.

7. Ripoll-Bosch, R.; Ripoll, G.; Álvarez-Rodríguez, J.; Blasco, I.; Panea, B.; Joy, M. Efecto del sexo y la explotación sobre la calidad de la canal y de la carne del cordero lechal de raza ojinegra. *ITEA Inf. Tec. Econ. Agrar.* **2012**, *108*, 522–536.
8. MARM. Estudio de Mercado Observatorio del Consumo y la Distribución Distribución Alimentaria. Available online: http://www.mapama.gob.es/es/alimentacion/temas/consumo-y-comercializacion-y-distribucion-alimentaria/informe_10_tcm30-128614.pdf (accessed on 21 May 2018).
9. Grunert, K. Future trends and consumer lifestyles with regard to meat consumption. *Meat Sci.* **2006**, *74*, 149–160. [CrossRef] [PubMed]
10. Cotes, A. *Modelos de Comportamiento del Consumidor de Productos Alimenticios Con Valor Agregado*; Universidad de Salamanca: Salamanca, Spain, 2010.
11. Gracia, A. Consumers' attitudes towards designation of origin lamb meat: Segmentation and profiles. *ITEA Inf. Tec. Econ. Agrar.* **2005**, *101*, 25–44.
12. Arcia, P.L. *Influencia de Las Caracteristicas Sensoriales y la Información Nutricional en la Respuesta de Los Consumidores a Alimentos Funcionales*; CSIC: Valencia, Spain, 2012.
13. Gracia, A. *Comportamiento del Consumidor de Carne de Cordero con Igp en Aragón*; AgEcon Search: Zaragoza, Spain, 2005; p. 132.
14. Bredahl, L.; Grunert, K.G. Identificación de los estilos de vida alimenticios en españa. *Investig. Agrar. Econ.* **1997**, *12*, 247–263.
15. Buckley, M.; Cowan, C.; McCarthy, M. The convenience food market in Great Britain: Convenience food lifestyle (CFL) segments. *Appetite* **2007**, *49*, 600–617. [CrossRef] [PubMed]
16. Grunert, K.G.; Perrea, T.; Zhou, Y.; Huang, G.; Sorensen, B.T.; Krystallis, A. Is food-related lifestyle (FRL) able to reveal food consumption patterns in non-western cultural environments? Its adaptation and application in urban china. *Appetite* **2011**, *56*, 357–367. [CrossRef] [PubMed]
17. Nie, C.; Zepeda, L. Lifestyle segmentation of us food shoppers to examine organic and local food consumption. *Appetite* **2011**, *57*, 28–37. [CrossRef] [PubMed]
18. Good, L.K.; Huddleston, P. Ethnocentrism of polish and Russian consumers: Are feelings and intentions related. *Int. Mark. Rev.* **1995**, *12*, 35–48. [CrossRef]
19. Sharma, S.; Shimp, T.A.; Shin, J. Consumer ethnocentrism: A test of antecedents and moderators. *J. Acad. Mark. Sci.* **1995**, *23*, 26–37. [CrossRef]
20. Worsley, A.; Skrzypiec, G. Teenage vegetarianism: Prevalence, social and cognitive contexts. *Appetite* **1998**, *30*, 151–170. [CrossRef] [PubMed]
21. Forestell, C.A.; Spaeth, A.M.; Kane, S.A. To eat or not to eat red meat. A closer look at the relationship between restrained eating and vegetarianism in college females. *Appetite* **2012**, *58*, 319–325. [CrossRef] [PubMed]
22. Larsson, C.L.; Klock, K.S.; Nordrehaug Åstrøm, A.; Haugejorden, O.; Johansson, G. Lifestyle-related characteristics of young low-meat consumers and omnivores in sweden and norway. *J. Adolesc. Health* **2002**, *31*, 190–198. [CrossRef]
23. Bello, L.; Calvo, D. The importance of intrinsic and extrinsic cues to expected and experienced quality: An empirical application for beef. *Food Qual. Prefer.* **2000**, *11*, 229–238. [CrossRef]
24. Dransfield, E.; Ngapo, T.M.; Nielsen, N.A.; Bredahl, L.; Sjödén, P.O.; Magnusson, M.; Campo, M.M.; Nute, G.R. Consumer choice and suggested price for pork as influenced by its appearance, taste and information concerning country of origin and organic pig production. *Meat Sci.* **2005**, *69*, 61–70. [CrossRef] [PubMed]
25. Bernabéu, R.; Tendero, A. Preference structure for lamb meat consumers. A Spanish case study. *Meat Sci.* **2005**, *71*, 464–470. [CrossRef] [PubMed]
26. Panea, B.; Ripoll, G.; Campo, M.M.; Albertí, P.; Bernués, A. Percepción del Consumidor Aragonés Sobre los Métodos de Producción y la Calidad Intrinseca y Extrínseca de la Carne de Cordero. In *XXXVIII Congreso Nacional de la Sociedad Española de Ovinotecnia y Caprinotecnia*; SEOC: Málaga, Spain, 2013; pp. 322–328.
27. Dodds, W.B.; Monroe, K.B.; Grewal, D. Effects of price, brand, and store information on buyers' product evaluations. *J. Mark. Res.* **1991**, *28*, 307–319. [CrossRef]
28. Banovic, M.; Grunert, K.G.; Barreira, M.M.; Fontes, M.A. Beef quality perception at the point of purchase: A study from portugal. *Food Qual. Prefer.* **2009**, *20*, 335–342. [CrossRef]
29. Steenkamp, J.B.E.M.; Van Trijp, H.C.M. Quality guidance: A consumer-based approach to food quality improvement using partial least squares. *Eur. Rev. Agric. Econ.* **1996**, *23*, 195–215. [CrossRef]

30. Verbeke, W.; Vackier, I. Profile and effects of consumer involvement in fresh meat. *Meat Sci.* **2004**, *67*, 159–168. [CrossRef] [PubMed]
31. Xue, H.; Mainville, D.; You, W.; Nayga, R.M., Jr. Consumer preferences and willingness to pay for grass-fed beef: Empirical evidence from in-store experiments. *Food Qual. Prefer.* **2010**, *21*, 857–866. [CrossRef]
32. Gil, J.M.; Gracia, A.; Sánchez, M. Market segmentation and willingness to pay for organic products in Spain. *Int. Food Agribus. Manag. Rev.* **2000**, *3*, 207–226. [CrossRef]

Article

Improving Cull Cow Meat Quality Using Vacuum Impregnation

Martha Y. Leal-Ramos [1,*], Alma D. Alarcón-Rojo [2], Néstor Gutiérrez-Méndez [1], Hugo Mújica-Paz [3], Felipe Rodríguez-Almeida [2] and Armando Quintero-Ramos [1]

1 Facultad de Ciencias Químicas, Universidad Autónoma de Chihuahua, 31125 Chihuahua, Mexico; ngutierrez@uach.mx (N.G.-M.); aquinter@uach.mx (A.Q.-R.)
2 Facultad de Zootecnia y Ecología, Universidad Autónoma de Chihuahua, 33820 Chihuahua, Mexico; aalarcon@uach.mx (A.D.A.-R.); frodrigu@uach.mx (F.R.-A.)
3 Departamento de Biotecnología e Ingeniería de Alimentos, Instituto Tecnológico y de Estudios Superiores de Monterrey, 64849 Monterrey, Mexico; h.mujica@itesm.mx
* Correspondence: yarelyleal@hotmail.com; Tel.: +52(614)236-6000

Received: 26 March 2018; Accepted: 30 April 2018; Published: 7 May 2018

Abstract: Boneless strip loins from mature cows (50 to 70 months of age) were vacuum impregnated (VI) with an isotonic solution (IS) of sodium chloride. This study sought to determine the vacuum impregnation and microstructural properties of meat from cull cows. The experiments were conducted by varying the pressure, p_1 (20.3, 71.1 kPa), and time, t_1 (0.5, 2.0, 4.0 h), of impregnation. After the VI step, the meat was kept for a time, t_2 (0.0, 0.5, 2.0, 4.0 h), in the IS under atmospheric pressure. The microstructural changes, impregnation, deformation, and porosity of the meat were measured in all the treatments. Impregnation and deformation levels in terms of volume fractions of the initial sample at the end of the vacuum step and the VI processes were calculated according to the mathematical model for deformation-relaxation and hydrodynamic mechanisms. Scanning electron microscopy (SEM) was used to study the microstructure of the vacuum-impregnated meat samples. Results showed that both the vacuum and atmospheric pressures generated a positive impregnation and deformation. The highest values of impregnation X (10.5%) and deformation γ (9.3%) were obtained at p_1 of 71.1 kPa and t_1 of 4.0 h. The sample effective porosity (ε_e) exhibited a significant interaction ($p < 0.01$) between $p_1 \times t_1$. The highest ε_e (14.0%) was achieved at p_1 of 20.3 kPa and t_1 of 4.0 h, whereas the most extended distension of meat fibers (98 μm) was observed at the highest levels of p_1, t_1, and t_2. These results indicate that meat from mature cows can undergo a vacuum-wetting process successfully, with an IS of sodium chloride to improve its quality.

Keywords: vacuum impregnation; sodium chloride brine; cull cows; meat quality; microstructure; moisture-enhanced meat

1. Introduction

In comparison to meat from young animals, beef from mature cows is usually tougher and less juicy [1]. As a result, 56.4% of mature cow meat is merchandised as beef trim for grinding and processing. The remaining 43.6% is sold at a lower price at the primal/subprimal level and as steaks/roasts in supermarkets and food service operations [2]. On the other hand, moisture enhancement is a value-adding meat processing technology widely used by the meat industry. This technology improves tenderness, juiciness, flavor, and consistency of whole-muscle meat products of reduced eating quality [3,4]. Moisture enhancement consists of adding a solution containing water, sodium chloride (NaCl), sodium phosphates, and sodium lactate as preservative into raw whole-muscle meat [3]. Salt and phosphates increase the water-holding capacity (WHC) of post-rigor meat by significantly altering its microstructure. Most of the water in muscle (95%) is held by capillary forces in the spaces arising from the arrangement of the thick filaments of myosin and the thin filaments of actin/tropomyosin within the myofibrils [5,6]. The introduction of salts into muscle tissue changes the number of charges on both thick and thin filaments, resulting in the repulsion and enlargement of the spaces between actin and myosin, which allows the incorporation of more water. When a greater amount of water is bound to the meat protein matrix, it becomes more swollen, softer, and juicier [5].

Moisture-enhanced meat (MEM) is prepared by injecting brine into sub-primal pieces (without further tumbling treatment), which are then portioned into steaks or small pieces of meat [3,4]. Unfortunately, current methods for brine injection in meat have significant disadvantages; even when brine is injected from multiple needles, as is common in commercial practice, the brine is not uniformly distributed throughout the meat. This uneven distribution of brine causes striping or streaking which can be seen as "light" and "dark" stripes over the MEM's surface, negatively impacting its appearance and overall quality [7,8]. Several studies have investigated novel techniques to overcome these problems and to accelerate mass transfer in whole-muscle meat-brine systems. The methods studied include ultrasound [9–12], high pressure [13–15], and vacuum impregnation (VI). Techniques involving VI have been used to reduce salting time and to increase water retention of dry-cured ham [16], tasajo [17], poultry [18], and salmon [19,20].

Vacuum impregnation allows the direct insertion of an external solution into a product through its pores in a fast, controlled, and uniform way without destroying its original structure [21]. Vacuum impregnation is conducted using consecutive vacuum and atmospheric pressure stages, which allow the impregnating solution to fill the food pores through deformation-relaxation and hydrodynamic mechanisms (HDM) [22]. The effect of VI on physical and structural properties of fruits has been studied in several products through the introduction of hypotonic, isotonic, and hypertonic solutions in plant tissue [23]. In the case of meat tissue, salts (sodium chloride and phosphates) by themselves can affect the microstructure of meat, specifically, the size of spaces between myosin and actin myofilaments that constitute the capillary pores of the meat. Nevertheless, it has been reported that pork meat that was treated in a brine of 0.005 g/cm^3 NaCl showed slight differences in comparison with the untreated sample, while the essential structure of the myofibrils appeared to be intact [24].

Factors affecting the effectiveness of VI on food matrices include the following: (1) flow properties of the external liquid; (2) operation variables such as compression ratio ($r \approx p_2/p_1$), where p_1 is the vacuum pressure applied to the solid food-liquid system, and p_2 is the atmospheric pressure, lengths of the vacuum (t_1), and atmospheric (t_2) pressure steps, temperature, etc.; and (3) mechanical and structural properties of the food solid matrix (porosity, pore size and shape, type of fluid-like gas or liquid occupying the pores, viscoelastic character, etc.) [25]. Vacuum impregnation properties of food are characteristic parameters which show the feasibility of liquid penetration during VI processes. The primary VI properties of a food including the following: sample volume fraction impregnated by the external liquid at the end of the vacuum step (X_1) and at the end of the VI process (X); sample volume deformation at the end of the vacuum step (γ_1) and at the end of the VI process (γ); the sample effective porosity (ε_e).

Currently, no information is available on the VI properties of meat from cull cows or its related microstructural changes. The aim of this work was to study the effects of the VI operation variables p_1, t_1, and t_2 on the sample impregnation and deformation levels at the end of the vacuum step (X_1 and γ_1) and at the end of the VI process (X and γ). It also aimed to examine VI effects on both sample effective porosity (ε_e) and related microstructural changes of meat from mature cows vacuum impregnated with an isotonic solution of sodium chloride to improve the quality of cull cow meat.

2. Materials and Methods

2.1. Raw Material and Sample Preparation

Boneless strip loins (IMPS #180) [26] of five cull cows (one loin from each cow) of varying breeds (Brangus, Charolais, Holstein, and unknown crossbreeds) and varying ages (approximately 50 to

70 months of age) were used. One loin from each Brangus, Charolais, and Holstein, and two loins from unknown crossbreds, respectively, were used. Loins were obtained from a Federal Inspection Type (TIF) slaughter plant, in Chihuahua, Chih., Mexico (28°38′07″ N, 106°05′20″ W), at 24 h after slaughter (refrigerated at 4 °C). The TIF plants are regulated based on the guidelines of the Mexican Ministry of Agriculture, Livestock, Rural Development, Fisheries and Food and meet the sanitary standards imposed by the United States Department of Agriculture (USDA). In the TIF abattoir, after being stunned by a captive bolt (model USSS-1 JARVIS®), the cattle were slaughtered, suspended by a hind leg, bled, and transferred to the production line to begin the process of removing the head, feet, skin, viscera, and the quartering of the carcass.

Loins were transported to the laboratory and stored at 2 °C for no more than five days. Beef strip loins were trimmed of excess fat and connective tissues. Each strip loin was cut into 50 sections (6 × 5 × 2 cm) with the slab's length perpendicular to the main loin axis.

2.2. Isotonic Solution Preparation

An isotonic solution (IS) was used in the VI experiments to avoid cell turgor alterations in beef. The IS was formulated with distilled water and sodium chloride (analytical grade). Water activity (a_w) of the IS was adjusted to that of fresh meat using the Favetto and Chirife equation (Equation (1)) [22], where k was the sodium chloride constant (0.0371), and m was the sodium chloride concentration (mol/kg water). The water activity of IS was confirmed with a hygrometer (Novasina IC-500AW-Lab, Talstrasse, Switzerland).

$$a_w = 1 - \mathrm{k}m \quad (1)$$

2.3. Analytical Methods

Representative samples taken from each loin were ground and homogenized separately. Homogenates were used for moisture, fat, protein, and ash content analysis (AOAC Official Methods no. 950.46, 991.36, 981.10, and 920.153, respectively [27]). Meat homogenates (1 g) were manually mixed with 9 mL of distilled water for 10 s. The pH of the mixtures was measured with a pH-meter (HANNA Instruments Model HI 255, Cluj-Napoca, Rumania) calibrated with standard buffer solutions at pH 4.0 and pH 7.0. Water activity (a_w) of fresh meat was determined with a hygrometer (Novasina IC-500AW-Lab, Talstrasse, Switzerland). The pycnometer method was used to determine the densities of IS (ρ_{IS}) at 2 ± 1 °C and paraffin wax (ρ_{wax}). All analyses were performed in triplicate.

For each section of meat (24 sections by each strip loin), the initial mass (M_0) was determined. Sample mass at the end of the vacuum step (M_1) or sample mass at the end of the VI process (M) was determined as appropriate. Initial volume (V_0) for each section of meat was measured using the buoyant force method based on Archimedes' principle [28]. A vessel containing a specific volume of IS at 2 ± 1 °C was placed on the scale plate and tared to zero. After its initial weight determination, each section of fresh meat was individually suspended by a nylon thread (0.05 mm thickness and 15 cm length) and wholly immersed in the IS so that the sample did not touch the walls or the bottom of the container. The individual immersion of each meat sample was so rapid (less than 2 min on average), that no significant changes in the temperature of the IS were observed when measured with a T-type thermocouple thermometer (Hanna Instruments Model HI 935004P, Cluj-Napoca, Rumania). The mass of liquid displaced by the sample (M_{B}) was recorded. Diameter and mass values of nylon thread were much smaller than those of meat sample; for this reason, these were neglected. The initial volume and apparent density of fresh meat ρ_{meat} (the density of meat including all remaining pores) were calculated by Equations (2) and (3), respectively.

$$V_0 = \frac{M_{\mathrm{B}}}{\rho_{\mathrm{IS}}} \quad (2)$$

$$\rho_{\mathrm{meat}} = \rho_{\mathrm{IS}} \cdot \frac{M_0}{M'_{\mathrm{B}}} \quad (3)$$

For measuring sample volume at the end of the vacuum step (V_1), each section of meat was coated with a thin layer of paraffin wax. The wax layer avoided the mass transfer between sample and liquid. Each wax-coated sample was weighed and wholly immersed again in the IS as mentioned previously. The mass of liquid displaced by the wax-coated sample (M'_B) was recorded. The mass of the wax layer (M_{wax}) was obtained by subtracting M_1 from the mass of the wax-covered meat sample. Sample volume at the end of the vacuum step was calculated by Equation (4).

$$V_1 = \frac{M'_B}{\rho_{IS}} - \frac{M_{wax}}{\rho_{wax}} \quad (4)$$

To measure the sample volume at the end of the vacuum impregnation process (V), each sample was coated with a thin layer of paraffin wax and treated like the sample at the end of the vacuum step. Five sections of meat by treatment were used.

2.4. Vacuum Impregnation Experiments

Vacuum impregnation treatments were carried out in a vacuum system. The vacuum system comprised a glass vacuum desiccator (Wheaton Dry-Seal, capacity 5 L) containing IS at 2 ± 1 °C and a vacuum pump (Felisa pump, Mod. 1600 L, México), both connected through a silicone vacuum hose. From a beef strip loin, 24 sections of meat were taken and assigned to six desiccators (four pieces per desiccator). Meat sections were immersed in the IS per the ratio meat: IS of 1:5 (w/w). The vacuum pressure p_1 (20.3 or 71.1 kPa) was applied in the desiccator for a length of vacuum step t_1 (0.5, 2.0 or 4.0 h). Vacuum pressures used in this work were in the maximum and minimum range reported by others for similar materials [18]. At the end of the first vacuum step, atmospheric pressure (p_2) was restored while samples remained immersed for a time t_2 (0.0, 0.5, 2.0, or 4.0 h). For each meat section, the impregnation X_1 or X (m^3 IS/m^3 initial sample) and deformation γ_1 or γ (m^3/m^3 initial sample) levels in terms of volume fractions of the initial sample (Equations (5)–(9)) at the end of vacuum step and VI processes were calculated. It should be noted that the X values represent the actual impregnation given by $X_1 + X_2$, where X_2 is the sample volume fraction impregnated by the external liquid during the atmospheric pressure step. Likewise, γ values represent the actual deformation given by $\gamma_1 + \gamma_2$, where γ_2 is the sample volume deformed during the atmospheric pressure step.

For each meat section that was vacuum-impregnated, the effective or initial porosity (ε_e) was calculated according to the mathematical model (Equation (9)) reported by (Fito et al., 1996) [29], using these parameters and a compression ratio of $r \approx p_2/p_{1abs}$ that considers the absolute vacuum pressure as $p_{1abs} = p_2 - p_1$. In these equations $(M_1 - M_0)$ and $(M - M_0)$ represented the masses of IS impregnated in the samples at the end of the vacuum step and the end of the VI process, respectively.

$$X_1 = \frac{M_1 - M_0}{\rho_{IS} V_0} \quad (5)$$

$$\gamma_1 = \frac{V_1 - V_0}{V_0} \quad (6)$$

$$X = \frac{M - M_0}{\rho_{IS} V_0} \quad (7)$$

$$\gamma = \frac{V - V_0}{V_0} \quad (8)$$

$$\varepsilon_e = \frac{(X - \gamma)r + \gamma_1}{r - 1} \quad (9)$$

2.5. Scanning Electron Microscopy

Scanning electron microscopy (SEM) was used to study the microstructure of the vacuum-impregnated meat samples. Preparation and analysis of meat samples for SEM was conducted using the method proposed by Dykstra [30]. Meat samples (fresh and vacuum-impregnated) of approximately 2 mm^3 were cut with a stainless steel surgical scalpel blade. Samples were fixed in 2.5% glutaraldehyde in 0.1 M (pH 7.3) sodium cacodylate buffer for 12 h at 4 °C. Afterward, samples were washed twice in 0.1 M sodium cacodylate (pH 7.3). Samples were then dehydrated in a graded series of 55, 75, 85, 95, and 100% ethanol (ethanol:water, v/v). Samples were placed in each dilution for 20 min. The final absolute ethanol treatment was repeated two times each for 20 min. The samples were ultra-dehydrated by the critical point with CO_2 (1100 psi, 31.1 °C) in a Samdri-780A, Tousimis Research Corporation critical point dryer (Rockville, MD, USA). The dry samples were then mounted on aluminum stubs with carbon conductive cement, coated with gold, and observed in a JEOL JSM-5800LV (JEOL, Freising, Germany) scanning electron microscope operating at an accelerating voltage of 10 kV.

2.6. Statistical Analysis

The experimental design was a randomized complete block in a 2 × 3 × 4 factorial arrangement of treatments: two vacuum pressures p_1 (20.3 and 71.1 kPa) × three vacuum step lengths t_1 (0.5, 2.0 and 4.0 h) × four atmospheric step lengths t_2 (0.0, 0.5, 2.0, and 4.0 h). Each beef strip loin from cull cows represented a block for a total of five blocks. The mixed procedure of SAS (SAS^{TM} version 9.2) was used to perform type III tests of fixed effects. The model used for the sample impregnation (X), deformation (γ), and ε_e data included p_1, t_1, t_2 and their interactions as fixed effects and block as a random effect. To test the trends in the effects of the lengths of the vacuum (t_1) and atmospheric (t_2) pressure steps when they were significant either as interactions with p_1 or as main effects without interaction effect, these effects were fitted as continuous rather than qualifying factors. First and second order regression coefficients were used for t_1 and first, second, and third order regression coefficients for t_2, under the complete model previously described and adjusted with a mixed procedure of SAS^{TM}. Regression coefficients of the highest order that were not significant were eliminated from the model.

3. Results and Discussion

3.1. Analytical Results

The proximate composition, pH, water activity, and apparent density values of fresh meat from cull cows are presented in Table 1. Moisture, protein, fat, ash, and pH of the meat were similar to those reported by Buford et al. [31] and Patten et al. [32] for meat from mature cows. Water activity of fresh meat from cull cows (0.98) was similar to that reported for beef (0.98 to 0.99) by Lewicki [33]. Meat pH ranged from 5.62 to 5.97. It has been widely reported that muscle pH is dependent on the amount of glycogen present in the muscles at the time of slaughter. Muscle glycogen content may be influenced by the animal's diet and stress before slaughter. If glycogen stores are depleted before slaughter, the pH will decline slowly, and a higher than normal ultimate pH will occur [32]. The amount of water bound within the muscle tissue greatly depends on the space available between the myofilaments of actin and myosin. This space forms the capillary pores of muscle tissue in which the pH value plays a vital role [3]. The highest water-holding capacity of fresh meat occurs near the isoelectric point of main meat proteins at pH 5.2 or in a range between 5.5 to 5.9 [3].

The apparent density value of meat from cull cows was 1.07 ± 0.01 g/cm^3, whereas density values of IS and paraffin wax were 1.02 and 0.76 g/cm^3, respectively (Table 1). Hildrum et al. [34] stated that lean meat tissue has a consistent density of 1.07 to 1.08 g/cm^3. Cull cows are variable with regard to breed, size, age (from first calving dairy cows, culled for low production, to mature cows), fatness, and physiological and health status [35]. Consequently, as a result of their heterogeneity, considerable variability exists in the chemical composition, structural-mechanical properties, and quality of this meat.

Table 1. Physical-Chemical Characteristics of Meat, Isotonic Solution and Paraffin Wax.

Properties	*Mean* ± SD
Meat	
Moisture (g/100 g meat) *	75.5 ± 3.0
Protein (g/100 g meat) *	20.2 ± 0.4
Fat (g/100 g meat) *	3.5 ± 2.5
Ash (g/100 g meat) *	1.2 ± 0.0
pH	5.8 ± 0.2
Water activity	0.98 ± 0.0
ρ_{meat} (g/cm^3)	1.07 ± 0.0
Isotonic solution	
Water activity	0.98 ± 0.0
ρ_{IS} (g/cm^3)	1.02 ± 0.01
Paraffin wax	
ρ_{wax} (g/cm^3)	0.76 ± 0.01

SD: standard deviation; *: values on wet weight basis.

3.2. *Volume of Brine Impregnated in the Meat with Vacuum Impregnation*

The sample volume fraction impregnated by the IS at the end of the VI process (X) as a function of the time at which the system was under vacuum (t_1) and under atmospheric pressure (t_2) for both levels of vacuum pressure (p_1) are shown in Figure 1a,b, respectively. Part of X was impregnated during t_1, which corresponds to X_1 (Figure 1a). In all cases, positive values of X_1 were achieved. In the vacuum step, two opposite fluxes occurred in the pores of a food: (1) the expansion and partial outflow of the product's internal gas which was accompanied by the product pore's native liquid and (2) the capillary inflow of the external liquid as a function of the interfacial tension of the liquid and the diameter of pores [23,29,36]. An important part of the pores in the matrix of meat is occupied by a free liquid phase that contains a small or neglected gas phase volume entrapped within it [25]. This free liquid phase may be released from the matrix in the vacuum step. Chiralt et al. [25] demonstrated that, in Manchego curd (cylinders) with low porosity (like meat), the amount of liquid phase that can be released by applying vacuum (50 mbar) was 2.3% and 3.4% (expressed as curd volume fraction), respectively, before and after submitting curd to a VI operation with an isotonic solution. This low porosity causes approximately 3% of the curd volume to be replaced by brine in the brine vacuum impregnation process, although the curd volume fraction occupied by the gas is minute (less than 1%). Positive values of X_1 could indicate that the capillary solution gain of the meat samples at vacuum pressure dominated over the losses of native liquid.

Salvatori et al. [21] reported a positive value of X_1 (0.9%) for mangoes, which was related to the capillary penetration of an external solution due to its fibrous structure. Both meat and mangoes have fibrous structures. Fito and Chiralt [23] reported that, if during VI process the $p_1 > 200$ mbar (20 kPa), then a great capillary contribution can be expected for products with a small pore diameter (≈10 µm or less). The spaces between the thick filaments of myosin and the thin filaments of actin-tropomyosin within myofibrils constitute the capillary pores where most of the water in meat is present [37]. Interfilament space has been observed to vary between 320 to 570 Å (0.03 to 0.06 µm) with pH, sarcomere length, ionic strength, osmotic pressure, and whether the muscle is pre- or post-rigor [37]. In viscoelastic matrices such as meat, pressure changes can promote sample deformations coupled with impregnation in the vacuum step where sample expansion occurs as well as in the atmospheric pressure step or compression period [25]. Values of sample volume fraction impregnated by the solution at the end of the vacuum step (Figure 1a) and at the end of the atmospheric pressure step (Figure 1b) are similar to the values of relative sample volume deformation at the end of the vacuum period (Figure 2a) as well as at the end of the atmospheric pressure period (Figure 2b). The above results indicate that deformation-relaxation phenomena dominate the hydrodynamic mechanism.

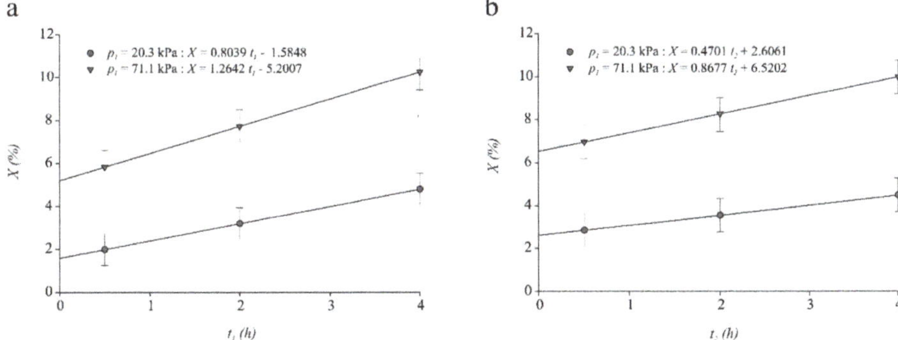

Figure 1. Mean ± Standard error (SE, n = 120) for meat sample volume fraction vacuum impregnated by the isotonic sodium chloride solution (X) for both levels of vacuum pressure (p_1) (mean ± SE): (**a**) as a function of the length of the vacuum pressure step (t_1) and, (**b**) as a function of the length of the atmospheric pressure step (t_2).

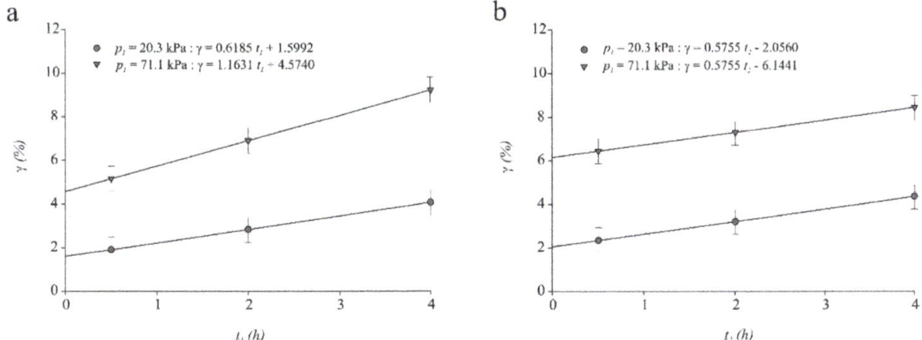

Figure 2. Mean ± SE (n = 120) for final meat sample volume deformation (γ) for both levels of vacuum pressure (p_1): (**a**) as a function of the length of the vacuum pressure step (t_1) and (**b**) as a function of the length of the atmospheric pressure step (t_2).

Meat sample volume fraction impregnated by the IS at the end of the VI process (X) exhibited significant ($p < 0.01$) $p_1 \times t_1$ and $p_1 \times t_2$ interaction effects, but no other interaction effects were important ($p < 0.05$). Figure 1a showed that the greater the vacuum level applied in the first step and the longer the vacuum step, the greater the average final impregnation degree ($p < 0.01$). The X was increased on average 0.80% ± 0.15 ($p < 0.01$) and 1.26% ± 0.15 ($p < 0.01$) by each hour that the system was under a p_1 of 20.3 kPa and 71.1 kPa, respectively (Figure 1a). Therefore, the highest average value of impregnation (10.4%) reached during the vacuum step was obtained at p_1 of 71.1 kPa applied by t_1 of 4.0 h (Figure 1a).

This observed phenomenon may have been a result of the high vacuum level, which allowed for better removal of native liquid from the meat pores. It also may also have resulted in a decrease in the remaining native liquid volume, with the same volume of external liquid, which was penetrated by HDM, as an effect of capillary pressure. Chiralt et al. [25] reported that the higher vacuum level p_1 in the system, the greater the capillary penetration in meat, fish, and cheese vacuum impregnated with a sodium chloride solution. On the other hand, the longer the vacuum step, the greater amount of native liquid left the meat structure, resulting in greater penetration of the external liquid.

Figure 1b shows that the average rates of change of X with respect to t_2 were 0.47% $\pm$ 0.17 ($p < 0.01$) and 0.87% $\pm$ 0.17 ($p < 0.01$) at 20.3 and 71.1 kPa, respectively ($p < 0.01$). It is worth mentioning that part of X impregnated during t_2 corresponds to X_2. Therefore, the highest average value of X_2 (9.7%) was reached at t_2 of 4.0 h, when the vacuum pressure applied to the system was 71.1 kPa (Figure 1b). When external pressure was restituted in the system, greatly deformed matrices may have relaxed the mechanical energy stored in their elastic structural elements which was in line with a continuous impregnation of their pores. The mechanical relaxation level and subsequent sample volume recovery depended on the viscoelastic properties of the matrix; the greater the elastic character, the higher the volume recovery and the coupled impregnation.

Brines for 10%-injected meat contain approximately 3% phosphates, 6% salt, and 30% lactate [3]. Notably, the average volume of the isotonic sodium chloride solution which was vacuum-impregnated in the meat during the vacuum step (10.4%) was very similar to that for injected meat (10%). Thus, the coupling of the hydrodynamic mechanism with deformation-relaxation phenomena produced a weight gain in each piece of meat without using high levels of either salt or sodium-based phosphates and lactate. Intake of dietary sodium has been linked to hypertension and, consequently, an increased risk of cardiovascular disease. In developed countries, meat and meat products contribute 21% to the sodium intake [38]. A challenge for the meat industry is to produce reduced-sodium raw or cooked meat products that consumers can enjoy as part of an ongoing healthier diet and lifestyle. Vacuum impregnation could be an excellent alternative to achieve moisture-enhanced meat products with a similar weight gain to traditional injected meats but with lower levels of sodium.

3.3. Volume Deformation of Cull Cow Meat Subjected to Vacuum Impregnation

The sample volume deformation at the end of the VI process (γ) as a function of the lengths of the vacuum (t_1) and atmospheric (t_2) pressure steps for both levels of vacuum pressure (p_1) are shown in Figure 2a,b, respectively. Meat samples increased in volume during the vacuum pressure step, as evidenced by positive values of γ_1(Figure 2a). This change in volume could be related to the prevailing capillary penetration of the external liquid during the vacuum step which promoted distension of the solid matrix of meat. Meat sample volume deformation at the end of the VI process (γ) exhibited a significant ($p < 0.01$) $p_1 \times t_1$ interaction.

The γ was increased on average 0.62% $\pm$ 0.14 ($p < 0.01$) and 1.16% $\pm$ 0.13 ($p < 0.01$) by each hour that the system was under a p_1 of 20.3 kPa and 71.1 kPa, respectively ($p < 0.01$) (Figure 2a). The highest mean value of γ (9.3%) was obtained at p_1 of 71.1 kPa and t_1 for 4.0 h. This change could be explained by the fact that the application of a more intense and prolonged vacuum step during the vacuum impregnation process may lead to a better substitution of a part of the sample-free liquid phase by the external liquid. This, in turn, will assist in the overall increase of sample volume. On the other hand, the length of the atmospheric step (t_2) had a significant positive linear effect ($p < 0.01$) on γ for both levels of p_1 (Figure 2b). This could have been a result of the water added during the atmospheric step, which allowed for an increase in the overall volume of the meat sample.

Other factors affecting the sample volume deformation at the end of the vacuum impregnation process were the mechanical properties of the food. Chiralt et al. [25] stated that the mechanical relaxation level and subsequent sample volume recovery would depend on the viscoelastic properties of the food matrix; the greater the elastic character, the higher the volume recovery and coupled impregnation. The connective tissue of meat from old cull cows is characterized by a higher number of cross-links within the collagen molecules. Collagen, the main component of muscle connective tissue, greatly influences the mechanical properties of meat. Lepetit [39] stated that the elastic modulus of the collagenous fraction of connective tissues is approximately proportional to the total number of collagen cross-links present per volume of meat. Thus, the greatest elastic character of meat from older cull cows could explain the positive values of γ observed.

3.4. Changes in the Sample Effective Porosity of Meat

The effective porosity (ε_e) of the meat sample as a function of t_1 exhibited a significant $p_1 \times t_1$ interaction effect ($p < 0.01$), as shown Figure 3. The ε_e increased 2.59% ± 0.25 ($p < 0.01$) and 0.41% ± 0.23 ($p < 0.01$) at 20.3 and 71.1 kPa, respectively (Figure 3). The t_2 did not have a significant effect ($p > 0.98$) on ε_e. The highest mean value of ε_e (14.0%) was obtained at p_1 for 20.3 kPa and t_1 of 4.0 h. Effective porosity is expressed a priori as the percentage of sample volume initially occupied by the gases but is defined more precisely as the sample volume fraction available for the deformation-relaxation phenomena, coupled with the hydrodynamic mechanism as indicated in Equation (9) [40]. Thus, 14.0% of the meat sample volume was occupied by pores or capillaries that could be impregnated with an external solution. The capillaries consist of the spaces between the actin and myosin filaments. It has been established that a high vacuum level increases the porosity of most plant tissues as a result of a high expansion and release of the gas inside the pores of vegetables. Moreover, a high vacuum level allows for the better removal of native liquid from the tissue structure. In this way, after the restoration of atmospheric pressure, a greater volume is available for impregnation phenomena [41]. In this study, it was observed that the meat sample effective porosity behaved contrary to that observed in vegetable tissues. The highest value of p_1 applied for longer t_1 promoted a greater impregnation level at the end of vacuum pressure step by the capillary. When the atmospheric pressure was restored in the meat–IS system, a reduced volume was available for the impregnation by HDM, which was observed by smaller values of ε_e.

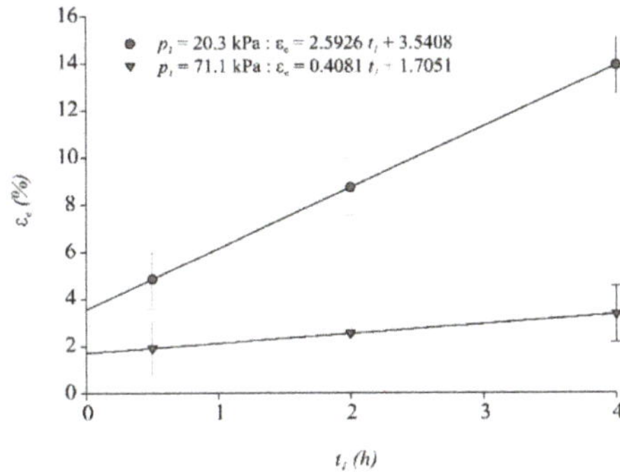

Figure 3. Mean ± SE (n = 120) for meat sample effective porosity (ε_e) as a function of the length of the vacuum step (t_1) for both levels of vacuum pressure (p_1).

3.5. Microstructural Analysis of Meat Vacuum Impregnated

The effect of vacuum impregnation on the microstructure of meat from cull cows is shown in Figure 4. As can be seen, meat sample vacuum impregnated at p_1 of 20.3 kPa, t_1 of 0.5 h and t_2 of 0.5 h (Figure 4b) showed slight swelling (approximately 50 μm) of muscle fibers with respect to that of fresh meat, whose essential structure of myofibrils appeared to remain intact (Figure 4a). Swelling of muscle fibers was more pronounced in the case of the vacuum-impregnated meat at the highest levels of factors studied. The largest thickness (98 μm) of muscle fibers was achieved at p_1 of 71.1 kPa, t_1 of 4.0 h, and t_2 of 4.0 h (Figure 4c). These results confirm the findings previously described with regard to the deformation properties of meat (Section 3.2). The influence of VI on meat microstructure could be explained by a more complete and homogeneous interchange of the gas and native liquid occupying the open pores (intermyofibrillar spaces) of the meat by the external isotonic sodium chloride solution. This results in a more complete contact between sodium chloride and meat proteins which leads to a more pronounced swelling of the meat matrix.

Graiver et al. [24] found that pork meat treated in a brine of 5 g/L NaCl showed slight differences to an untreated sample; however, the essential structure of the myofibrils appeared to remain intact. Fibers immersed in a NaCl solution of 140 g/L showed swelling as a result of salt ions diminishing the attractive forces between adjacent protein molecules. This resulted in an enlargement of the distance between the proteins and the meat which underwent swelling. Meat treated in a brine of NaCl 330 g/L produced fragmented and dehydrated fibers with a granular appearance. In the present study, 18.95 g/L brine concentration was applied. Therefore, one may assume that microstructure changes were predominantly caused by the vacuum impregnation treatment as opposed to the effects of salt.

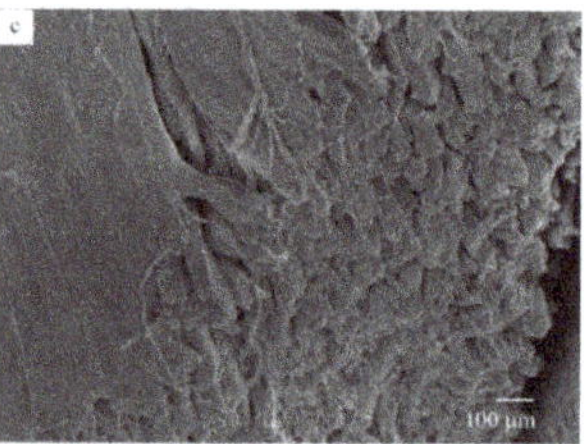

Figure 4. Scanning electron microscopy (SEM) micrographs of (**a**) fresh meat from cull cows; (**b**) vacuum-impregnated meat at p_1 of 20.3 kPa, t_1 of 0.5 h, and t_2 of 0.5 h, and (**c**) vacuum-impregnated meat at p_1 of 71.1 kPa, t_1 of 4.0 h, and t_2 of 4.0 h at 80× magnification.

4. Conclusions

For all vacuum impregnation conditions studied, positive values of impregnation (X_1 and X) and deformation (γ_1 and γ) were obtained. The coupling of a hydrodynamic mechanism with deformation-relaxation phenomena during a vacuum impregnation process with an isotonic sodium chloride solution resulted in a brine gain in each piece of meat. This brine gain was similar to the most common level of injected solution in moisture-enhanced meat (10%) without using brine injection or high levels of salt and sodium phosphates which increase sodium levels in meat. Contrary to processes in many plant tissues, the highest value of p_1 applied for longer t_1 promoted a greater impregnation level at the end of vacuum pressure step by capillary. This resulted in a diminished volume of meat samples available for impregnation by HDM when atmospheric pressure was restored in the meat–IS system. Accordingly, there were smaller values of ε_e as p_1 increased and t_1 decreased.

Vacuum impregnation was proven to be a useful technique in the addition of an external solution to small pieces of meat from cull cows. Vacuum impregnation may present a good alternative in the production of moisture-enhanced meat products, offering similar weight gain as that of traditional injected meat but with lower levels of sodium and a more complete distribution of the brine into meat. The amount and distribution of water inside raw meat has a considerable influence on its tenderness and juiciness after cooking. A high amount of intra-myofibrillar water is associated with more tender meat.

Author Contributions: M.Y.L.-R. and A.D.A.-R. conceived and designed the experiments; M.Y.L.-R. performed the experiments; M.Y.L.-R., N.G.-M. and F.R.-A. analyzed the data; A.D.A.-R., N.G.-M., H.M.-P. and A.Q.-R. contributed reagents/materials/analysis tools. All authors discussed the results and contributed to the final manuscript.

Conflicts of Interest: The authors declare no conflict of interest.

References

1. Hoffman, L. Sensory and physical characteristics of enhanced vs. Non-enhanced meat from mature cows. *Meat Sci.* **2006**, *72*, 195–202. [CrossRef] [PubMed]
2. Smith, D.P.; Acton, J.C. Marination, cooking, and curing of poultry products. In *Poultry Meat Processing*; Sams, A.R., Ed.; Taylor & Francis Group LLC: Boca Raton, FL, USA, 2001; pp. 257–280.

3. Feiner, G. *Meat Products Handbook Practical Science and Technology*; Publishing Limited: Cambridge, UK, 2006.
4. Vihavainen, E.J.; Björkroth, K.J. Spoilage of value-added, high-oxygen modified-atmosphere packaged raw beef steaks by leuconostoc gasicomitatum and leuconostoc gelidum. *Int. J. Food Microbiol.* **2007**, *119*, 340–345. [CrossRef] [PubMed]
5. Belitz, H.; Grosch, W.; Schieberle, P. *Food Chemistry*; Springer: Berlin, Germany, 2009.
6. Huff-Lonergan, E.; Lonergan, S.M. Mechanisms of water-holding capacity of meat: The role of postmortem biochemical and structural changes. *Meat Sci.* **2005**, *71*, 194–204. [CrossRef] [PubMed]
7. Uttaro, B.; Aalhus, J. Effect of thawing rate on distribution of an injected salt and phosphate brine in beef. *Meat Sci.* **2007**, *75*, 480–486. [CrossRef] [PubMed]
8. Uttaro, B.; Badoni, M.; Zawadski, S.; Gill, C. Effects of the pressure, flow rate and delivered volume of brine on the distributions of brine and bacteria in brine-injected meat. *Food Control* **2011**, *22*, 180–185. [CrossRef]
9. Cárcel, J.A.; Benedito, J.; Bon, J.; Mulet, A. High intensity ultrasound effects on meat brining. *Meat Sci.* **2007**, *76*, 611–619. [CrossRef] [PubMed]
10. Leal-Ramos, M.Y.; Alarcon-Rojo, A.D.; Mason, T.J.; Paniwnyk, L.; Alarjah, M. Ultrasound-enhanced mass transfer in halal compared with non-halal chicken. *J. Sci. Food Agric.* **2011**, *91*, 130–133. [CrossRef] [PubMed]
11. Siró, I.; Vén, C.; Balla, C.; Jónás, G.; Zeke, I.; Friedrich, L. Application of an ultrasonic assisted curing technique for improving the diffusion of sodium chloride in porcine meat. *J. Food Eng.* **2009**, *91*, 353–362. [CrossRef]
12. Stadnik, J.; Dolatowski, Z.J.; Baranowska, H.M. Effect of ultrasound treatment on water holding properties and microstructure of beef (m. Semimembranosus) during ageing. *LWT-Food Sci. Technol.* **2008**, *41*, 2151–2158. [CrossRef]
13. Lakshmanan, R.; Parkinson, J.A.; Piggott, J.R. High-pressure processing and water-holding capacity of fresh and cold-smoked salmon (*salmo salar*). *LWT-Food Sci. Technol.* **2007**, *40*, 544–551. [CrossRef]
14. Villacís, M.; Rastogi, N.; Balasubramaniam, V. Effect of high pressure on moisture and NaCl diffusion into turkey breast. *LWT-Food Sci. Technol.* **2008**, *41*, 836–844. [CrossRef]
15. Yashoda, K.; Rao, R.J.; Mahendrakar, N.; Rao, D.N. Marination of sheep muscles under pressure and its effect on meat texture quality. *J. Muscle Foods* **2005**, *16*, 184–191. [CrossRef]
16. Barat, J.M.; Grau, R.; Fito, P.; Chiralt, A. Vacuum salting treatment for the accelerated processing of dry-cured ham. In *Advanced Technologies for Meat Processing*; Nollet, L.M.L., Toldrá, F., Eds.; Taylor & Francis Group LLC: Boca Raton, FL, USA, 2006; pp. 353–370.
17. Barat, J.; Andujar, G.; Andrés, A.; Arguelles, A.; Fito, P. Salting studies during tasajo making. In *Osmotic Dehydration and Vacuum Impregnation Applications in Food Industries*; Fito, P., Chiralt, A., Barat, J.M., Spiess, W.E.L., Behsnilian, D., Eds.; Technomic Publishing Company Inc.: Lancaster, PA, USA, 2001; pp. 171–183.
18. Deumier, F.; Trystram, G.; Collignan, A.; Guédider, L.; Bohuon, P. Pulsed vacuum brining of poultry meat: Interpretation of mass transfer mechanisms. *J. Food Eng.* **2003**, *58*, 85–93. [CrossRef]
19. Bugueño, G.; Escriche, I.; Chiralt, A.; Pérez-Juan, M.; Serra, J.A.; Camacho, M.M. Use of vacuum impregnation in smoked salmon manufacturing. In *Osmotic Dehydration & Vacuum Impregnation Applications in Food Industries*; Fito, P., Chiralt, A., Barat, J.M., Spiess, W.E.L., Behsnilian, D., Eds.; Technomic Publishing Company, Inc.: Lancaster, PA, USA, 2001; pp. 195–206.
20. Larrazábal-Fuentes, M.J.; Escriche-Roberto, I.; Camacho-Vidal, M.d.M. Use of immersion and vacuum impregnation in marinated salmon (salmo salar) production. *J. Food Process. Preserv.* **2009**, *33*, 635–650. [CrossRef]
21. Salvatori, D.; Andres, A.; Chiralt, A.; Fito, P. The response of some properties of fruits to vacuum impregnation. *J. Food Process Eng.* **1998**, *21*, 59–73. [CrossRef]
22. Valdez-Fragoso, A.; Sáenz-Hernández, C.; Welti-Chanes, J.; Mújica-Paz, H. Cherry pepper pickling: Mass transport and firmness parameters and stability indicators. *J. Food Eng.* **2009**, *95*, 648–655. [CrossRef]
23. Fito, P.; Chiralt, A. Vacuum impregnation of plant tissues. In *Minimally Processed Fruits and Vegetables*; Alzamora, S.M., Tapia, M.S., López-Malo, A., Eds.; Aspen Publication Inc.: Frederick, MD, USA, 2000; pp. 189–201.
24. Graiver, N.; Pinotti, A.; Califano, A.; Zaritzky, N. Diffusion of sodium chloride in pork tissue. *J. Food Eng.* **2006**, *77*, 910–918. [CrossRef]

25. Chiralt, A.; Fito, P.; Barat, J.M.; Andrés, A.; González-Martínez, C.; Escriche, I.; Camacho, M.M. Use of vacuum impregnation in food salting process. *J. Food Eng.* **2001**, *49*, 141–151. [CrossRef]
26. USDA. *Institutional Meat Purchasing Specifications for Fresh Beef*; Serv, A.M., Ed.; USDA: Washington, DC, USA, 2014.
27. Cunnif, P. *Official Methods of Analysis of Aoac International*, 16th ed.; Association of Official Analytical Chemists International: Arlington, VA, USA, 1998.
28. Rahman, M.S. Mass-volume-area-related properties of foods. In *Engineering Properties of Foods*; Rao, M.A., Rizvi, S.S.H., Datta, A.K., Eds.; CRC Press Taylor & Francis Group: Boca Raton, FL, USA, 2005; pp. 1–40.
29. Fito, P.; Andrés, A.; Chiralt, A.; Pardo, P. Coupling of hydrodynamic mechanism and deformation-relaxation phenomena during vacuum treatments in solid porous food-liquid systems. *J. Food Eng.* **1996**, *27*, 229–240. [CrossRef]
30. Dykstra, M.J. *A Manual of Applied Techniques for Biological Electron Microscopy*; Springer Science & Business Media: New York, NY, USA, 1993; p. 78.
31. Buford, M.; Calkins, C.; Johnson, D.; Gwartney, B. *Cow Muscle Profiling: Processing Traits of 21 Muscles from Beef and Dairy Cow Carcasses*; University of Nebraska-Lincoln: Lincoln, NE, USA, 2003.
32. Patten, L.; Hodgen, J.; Stelzleni, A.; Calkins, C.; Johnson, D.; Gwartney, B. Chemical properties of cow and beef muscles: Benchmarking the differences and similarities 1 2. *J. Anim. Sci.* **2008**, *86*, 1904–1916. [CrossRef] [PubMed]
33. Lewicki, P.P. Data and models of water activity II: Solid foods. In *Food Properties Handbook*; CRC Press Taylor & Francis Group: Boca Raton, FL, USA, 2009.
34. Hildrum, K.I.; Wold, J.P.; Segtnan, V.H.; Renou, J.P.; Dufour, E. New spectroscopic techniques for online monitoring of meat quality. In *Advanced Technologies for Meat Processing*; Nollet, L.M.L., Toldrá, F., Eds.; CRC Press and Taylor & Francis Group: Boca Raton, FL, USA, 2006; pp. 88–123.
35. Malterre, C.; Jones, S.D.M. Meat production from heifers and cull cows. In *Beef Cattle Production (World Animal Science. C: Production-Approach 5)*; Jarrige, R., Beéranger, C., Eds.; Elsevier Scince Publishers: Holland, The Netherlands, 1992; pp. 357–375.
36. Andres, A.; Salvatori, D.; Albors, A.; Chiralt, A.; Fito, P. Vacuum impregnation viability of some fruits and vegetables. In *Osmotic Dehydration & Vacuum Impregnation: Aplications in Food Industries*; Fito, P., Chiralt, A., Barat, J.M., Spiess, W.E., Behsnilian, D., Eds.; Technomic Publishing Company Inc.: Lancaster, PA, USA, 2001; pp. 53–60.
37. Lawrie, R.A.; Ledward, D.A. *Lawries's Meat Science*; CRC Press LLC: Boca Raton, FL, USA, 2006.
38. Desmond, E. Reducing salt: A challenge for the meat industry. *Meat Sci.* **2006**, *74*, 188–196. [CrossRef] [PubMed]
39. Lepetit, J. A theoretical approach of the relationships between collagen content, collagen cross-links and meat tenderness. *Meat Sci.* **2007**, *76*, 147–159. [CrossRef] [PubMed]
40. Saurel, R. The use of vacuum technology to improve processed fruit and vegetables. In *Fruit and Vegetable Processing Improving Quality*; Jongen, W., Ed.; CRC Press LLC: Boca Raton, FL, USA, 2002; pp. 363–380.
41. Derossi, A.; De Pilli, T.; Severini, C. Reduction in the ph of vegetables by vacuum impregnation: A study on pepper. *J. Food Eng.* **2010**, *99*, 9–15. [CrossRef]

Article

Prevalence of Pathogens in Poultry Meat: A Meta-Analysis of European Published Surveys

Andiara Gonçalves-Tenório, Beatriz Nunes Silva, Vânia Rodrigues, Vasco Cadavez and Ursula Gonzales-Barron *

CIMO Mountain Research Centre, School of Agriculture, Polytechnic Institute of Bragança, Campus de Santa Apolónia, 5301-855 Bragança, Portugal; andiaragoncalves@gmail.com (A.G.-T.); beatriznsilvaa@gmail.com (B.N.S.); vania.rodrigues@ipb.pt (V.R.); vcadavez@ipb.pt (V.C.)

* Correspondence: ubarron@ipb.pt; Tel.: +351-273-303-325

Received: 12 April 2018; Accepted: 26 April 2018; Published: 3 May 2018

Abstract: The objective of this study was to investigate and summarize the levels of incidence of *Salmonella* spp., *Listeria monocytogenes*, *Staphylococcus aureus* and *Campylobacter* spp. in poultry meat commercialized in Europe. After systematic review, incidence data and study characteristics were extracted from 78 studies conducted in 21 European countries. Pooled prevalence values from 203 extracted observations were estimated from random-effects meta-analysis models adjusted by pathogen, poultry type, sampling stage, cold preservation type, meat cutting type and packaging status. The results suggest that *S. aureus* is the main pathogen detected in poultry meat (38.5%; 95% CI: 25.4–53.4), followed by *Campylobacter* spp. (33.3%; 95% CI: 22.3–46.4%), while *L. monocytogenes* and *Salmonella* spp. present lower prevalence (19.3%; 95% CI: 14.4–25.3% and 7.10%; 95% CI: 4.60–10.8%, respectively). Despite the differences in prevalence, all pathogens were found in chicken and other poultry meats, at both end-processing step and retail level, in packed and unpacked products and in several meat cutting types. Prevalence data on cold preservation products also revealed that chilling and freezing can reduce the proliferation of pathogens but might not be able to inactivate them. The results of this meta-analysis highlight that further risk management strategies are needed to reduce pathogen incidence in poultry meat throughout the entire food chain across Europe, in particular for *S. aureus* and *Campylobacter* spp.

Keywords: chicken; *Listeria monocytogenes*; *Campylobacter*; *Salmonella*; *Staphylococcus aureus*

---

1. Introduction

Among the group of infectious bacteria, *Salmonella* spp., *Listeria monocytogenes*, enterotoxigenic *Staphylococcus aureus* and *Campylobacter* spp. are the main contaminants in food due to their high occurrence worldwide and for being major causes of gastroenteritis in humans [1,2].

The food poisoning symptoms caused by *Salmonella* spp. and *Campylobacter* spp. are usually abdominal cramps, fever, vomiting and diarrhea, which is often bloody in the case of the latter pathogen [3]. Listeriosis is the foodborne disease more likely to lead to hospitalization and mortality, and symptoms can include headache, stiff neck, confusion, loss of balance and convulsions in addition to fever and muscle aches. In pregnant women, listeriosis is particularly dangerous as it can lead to miscarriage, stillbirth, premature delivery, or life-threatening infection of the new-born [3]. *S. aureus* food poisoning symptoms include diarrhea, nausea, stomach cramps and vomiting [3]. Although it is not regarded as a disease as severe as listeriosis, for instance, in some rare cases, acute staphylococcal enterotoxicosis can cause death due to complications [4].

Epidemiological studies have suggested that products of poultry origin are among the most common vehicles for transmission of *Salmonella* spp. and *Campylobacter* spp. [5,6]. According to the European Food Safety Authority (EFSA), there is ample evidence that *Campylobacter* spp. is a foodborne hazard related to poultry meat, by cross-contamination from contaminated broiler meat to ready-to-eat foods [7]. It is estimated that 20% to 30% of human cases of campylobacteriosis are caused by handling, preparation, and consumption of broiler meat, while 50% to 80% may be attributed to the chicken reservoir as a whole [8]. EFSA also points *Salmonella* spp. as a high priority pathogen regarding poultry meat inspection [8], as poultry meat and poultry products are common sources of sporadic and outbreak-related cases of salmonellosis [9].

Listeriosis is usually associated with ready-to-eat products, including products made of poultry meat, in which contamination has occurred before or during processing, followed by growth during prolonged refrigerated storage [8]. With regard to *S. aureus*, it is considered that the risk of disease is more related with improper hygiene and storage throughout the food chain than its occurrence in raw meat, as this bacterium can be found naturally in poultry meats but also in the environment [8], facilitating cross-contamination between surrounding areas and poultry products.

It is important to perform research on pathogens' detection in food to assess the main zoonotic disease carriers and to combine this information with other available in literature, through meta-analysis. By doing this, it is possible to address a wide range of food safety questions, such as the incidence of diseases and the risks imposed to public health [10]. Hence, the objective of this study was to conduct a systematic review of the incidence of *Salmonella* spp., *L. monocytogenes*, *S. aureus* and *Campylobacter* spp. in raw poultry meat at the end of the processing stage and sold at European retail establishments. In addition, meta-analysis was applied to quantitatively combine and compare the incidence of each pathogen: (i) in poultry meat as a whole; (ii) in chicken vs. other poultry meats; (iii) in poultry by different sampling stages; (iv) in chicken or other poultry meats according to distinct types of cut; (v) in chicken or other poultry meats by packaging status; and (vi) in poultry by cold preservation type.

2. Materials and Methods

As a statistical analysis of a large collection of results from published studies, meta-analysis aims to integrate and interpret the findings to achieve comprehensive conclusions that the individual studies alone would not demonstrate clearly [11]. In this study, the *population* is defined as raw poultry meat surveyed at the end of the processing stage or at retail establishments in Europe, while the *measured outcome* is the detection of pathogens. Electronic literature search was carried out in Scopus and Scielo databases to find articles and official reports published since 2000 summarizing the incidence of microbiological hazards in chicken meat produced and commercialized in Europe. The search was done systematically and aimed to find quality studies validated by the scientific community. Grey literature was not procured for two reasons; on the one hand, to avoid data validity concerns; and, on the other hand, to avoid data duplication, since it is probable that high-quality theses and reports be also published in peer-reviewed journals.

The bibliographic searches were undertaken using a formula that combined terms regarding the existence (prevalence, incidence, occurrence, quality, contamination, survey, sampling) of pathogens (*Campylobacter, Listeria monocytogenes, Salmonella, Staphylococcus aureus*) in the selected products (chicken, turkey, broiler, poultry), while excluding other meta-analysis studies, systematic reviews and articles regarding feed and where products were artificially inoculated (artificial, inocul*, spiked). All the terms were combined by properly applying the AND, OR and AND NOT logical connectors.

After assessing all the information from the recovered publications, seventy-eight primary studies [6,12–88] published from 2000 until April 2017 were considered appropriate for inclusion as they satisfied the following criteria: (i) reported outcomes from European processing units or retail establishments; (ii) use of approved microbiological methods; and (iii) presenting sufficient and extractable data (in presence/absence format).

As the incidence of microbial hazards in poultry meat is a binary trait (i.e., a sample tests either positive or negative for the pathogen), the parameter used to measure the effect size was the raw proportion p (calculated as the number of successes or positive samples, s, divided by the total sample size, n). Thus, from each primary study, the number of positive samples s and the total number n were extracted, and p was calculated. Additionally, study characteristics, such as country of origin, survey's year, sample weight (in g), poultry class (chicken or other poultry), type of meat cut (i.e., whole carcass, pre-cut, minced meat or giblets), sampling stage (i.e., end of processing or retail), packaging status (i.e., unpacked or packed in modified atmosphere packaging—MAP,) and cold preservation type (i.e., chilled or frozen) were annotated.

Several multilevel meta-analysis models were fitted to appropriate data subsets in order to estimate overall or pooled prevalence values for: (i) each pathogen in poultry meat as a whole; (ii) each pathogen in chicken and in other poultry meats; (iii) each pathogen in poultry meat disaggregated by sampling stage; (iv) each pathogen in chicken and in other poultry meats by type of cut; (v) each pathogen in chicken and in other poultry meats by packaging status; and (vi) each pathogen in poultry meat by different cold preservation types. For a detailed explanation on the calculation of study's weight (precision) and multilevel meta-analysis modelling for incidence data, refer to Xavier et al. [11] and Viechtbauer et al. [89]. Meta-analysis models, and Galbraith and forest plots were built in RStudio version 1.0.136 (RStudio, Boston, MA, USA) using the 'metafor' [89] and 'sqldf' [90] packages.

3. Results

Following study quality checking, a total of 203 observations of positive and negative results of incidence of foodborne pathogens in poultry meats were retrieved.

Most of the observations excerpted from the primary studies were regarding *L. monocytogenes* ($n = 75$), followed by *Salmonella* spp. ($n = 51$) and *Campylobacter* spp. ($n = 50$). *S. aureus* was the pathogen with the fewest observations retrieved ($n = 27$). It is worth mentioning that the meta-analysis results represent a synthesis of 21 European countries; namely, Austria, Bulgaria, Croatia, Denmark, Estonia, Finland, Germany, Greece, Hungary, Ireland, Italy, the Netherlands, Poland, Portugal, Romania, Serbia, Spain, Sweden, Switzerland, UK, and Turkey. Therefore, the pooled prevalence estimates, to be presented as follows, cannot be generalized to other European countries.

3.1. Incidence of Pathogens in Poultry Meat

The overall prevalence of the four pathogens in poultry meat can be read from the Galbraith plots presented in Figure 1. In a meta-analytical Galbraith plot, the standardized logit transformation of the incidence value (y-axis) taken from a study is plotted against its precision (x-axis), and the meta-analysis solution is shown on the prevalence scale to the right. None of the plots showed evidence of outliers, yet all of them suggest the presence of heterogeneity among the observations extracted from the literature, fact also corroborated by the intra-class correlation I^2 values, which in most cases were higher than 0.25 (Table 1). In meta-analysis, between-study variability can be considered as significant when it represents at least 25% of the total variability in the outcome measure.

Overall, *S. aureus* bears the highest pooled prevalence in poultry meat (38.5%; 95% CI: 25.4–53.4%), followed by *Campylobacter* spp. (33.3%; 95% CI: 22.3–46.4%) and *L. monocytogenes* (19.3%; 95% CI: 14.4–25.3%), while *Salmonella* spp. was the pathogen of lowest prevalence (7.10%; 95% CI: 4.60–10.8%).

Table 1. Meta-analysis of the incidence of pathogens in poultry meat by type of cut. Heterogeneity analysis comprises between-study variability to total variability ratio (I^2) and proportion of between-study variability explained by type of meat cut (R^2).

Microorganism	Product (*n*)	Pooled Prevalence	95% CI Pooled Prevalence	Heterogeneity
Campylobacter spp.	**Chicken (30)**	0.486	(0.281–0.696)	$I^2 = 0.419$ $R^2 = 0.251$
	Pre-cut (17)	0.305	(0.115–0.595)	
	Whole carcass (11)	0.719	(0.432–0.896)	
	Other Poultry (20)	0.230	(0.090–0.472)	$I^2 = 0.211$ $R^2 = 0.204$
	Pre cut (10)	0.248	(0.103–0.487)	
	Whole carcass (8)	0.311	(0.138–0.559)	

Table 1. *Cont.*

Microorganism	Product (*n*)	Pooled Prevalence	95% CI Pooled Prevalence	Heterogeneity
L. monocytogenes	**Chicken (45)**	0.210	(0.141–0.301)	$I^2 = 0.330$ $R^2 = 0.046$
	Pre-cut (33)	0.202	(0.125–0.311)	
	Whole carcass (9)	0.246	(0.131–0.414)	
	Other Poultry (30)	0.129	(0.078–0.204)	$I^2 = 0.408$ $R^2 = 0.012$
	Pre-cut (16)	0.117	(0.063–0.207)	
	Whole carcass (9)	0.143	(0.058–0.310)	
	Mince (5)	0.190	(0.073–0.409)	
Salmonella spp.	**Chicken (26)**	0.032	(0.016–0.064)	$I^2 = 0.455$ $R^2 = 0.154$
	Pre-cut (18)	0.058	(0.027–0.122)	
	Whole carcass (5)	0.023	(0.007–0.068)	
	Giblets (3)	0.015	(0.003–0.059)	
	Other Poultry (25)	0.057	(0.027–0.119)	$I^2 = 0.285$ $R^2 = 0.176$
	Pre-cut (17)	0.058	(0.026–0.125)	
	Whole carcass (4)	0.141	(0.049–0.343)	
	Mince (4)	0.119	(0.032–0.357)	
S. aureus	**Chicken (16)**	0.399	(0.198–0.640)	$I^2 = 0.483$ $R^2 = 0.246$
	Pre-cut (7)	0.283	(0.090–0.612)	
	Whole carcass (3)	0.726	(0.386–0.918)	
	Giblets (6)	0.430	(0.169–0.736)	
	Other Poultry (11)	0.431	(0.204–0.692)	$I^2 = 0.313$ $R^2 = 0.157$
	Pre-cut (6)	0.298	(0.151–0.502)	
	Whole carcass (5)	0.413	(0.189–0.679)	

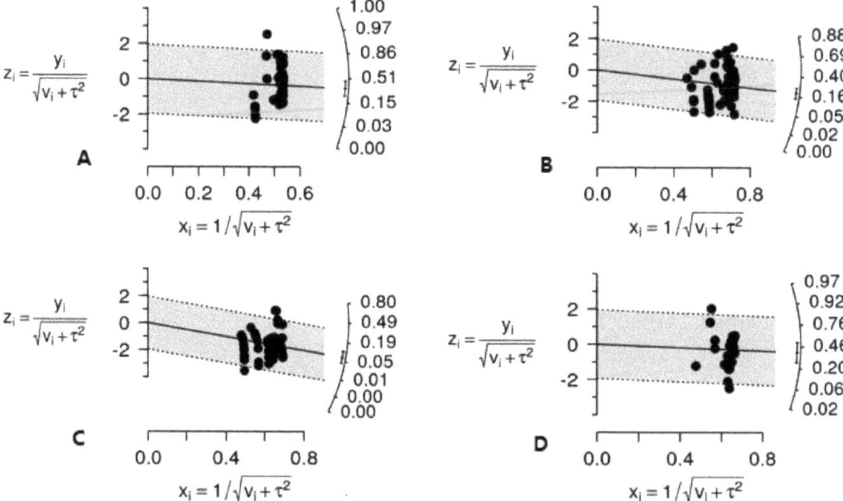

Figure 1. Galbraith plots of the meta-analyses of the incidence of the four pathogens in poultry meat surveyed at the end-processing stage and at retail establishments in Europe; *Campylobacter* spp. overall prevalence 33.3%; 95% CI: 22.3–46.4% (**A**); *L. monocytogenes* overall prevalence 19.3%; 95% CI: 14.4–25.3% (**B**); *Salmonella* spp. overall prevalence 7.10%; 95% CI: 4.60–10.8% (**C**); *S. aureus* overall prevalence 38.5%; 95% CI: 25.4–53.4% (**D**).For each observation i, y_i denotes the logit-transformed incidence; v_i the sampling variance; τ^2 the between-study variance; x_i the incidence value's precision; and z_i, the standardized logit-transformed incidence

3.2. Incidence of Pathogens by Poultry Type: Chicken Vs. Other Poultry Meats

The meta-analysis results on the prevalence of pathogens in poultry meat by poultry class are presented in Table 1. In chicken meat, *Campylobacter* spp. presented the highest prevalence (48.6%; 95% CI: 28.1–69.6%), followed by *S. aureus* (39.9%; 95% CI: 19.8–64.0%). For other poultry meats, such as turkey and duck, the microorganism of highest prevalence was *S. aureus* (43.1%; 95% CI: 20.4–69.2%), followed by *Campylobacter* spp. (23.0%; 95% CI: 9.00–47.2%). Despite the outcomes of this meta-analysis are, in principle, not directly comparable to the official incidence figures from the latest EFSA report [91]—since the former involves a precision-weighted average of results spanning from 2000 to 2017, while the latter compiles non-weighted averages from 2016—the EFSA's high mean incidence of *Campylobacter* in broiler meat (36.7%) and turkey meat (11.0%) [91] supported our findings. The pooled prevalence estimates for *Salmonella* spp. in chicken meat (3.2%; 95% CI: 1.6–6.4%) and other poultry meats (5.7%; 95% CI: 2.7–11.9%) were lower than the mean incidence values of *Salmonella* spp. in fresh broiler meat (6.4%) and fresh turkey meat (7.7%) from the 2016 EFSA report [91], although such values were still within the confidence intervals of the meta-analytical estimates. The pooled prevalence estimates of *S. aureus* in chicken/other poultry and of *L. monocytogenes* in chicken (21.0%; 95% CI: 14.1–30.1%) and other poultry meats (12.9%; 95% CI: 7.80–20.4%) could not be contrasted against EU official figures since neither *L. monocytogenes* nor *S. aureus* in raw meats or meat preparations are of compulsory notification to EFSA.

3.3. Incidence of Pathogens by Type of Cut

For different types of cuts commercialized (pre-cut, whole carcass, giblets, mince), it was observed that, in whole carcasses, *S. aureus* is the pathogen of greatest concern as it has the highest prevalence (72.6%; 95% CI: 38.6–91.8%) while in pre-cut meat, *Campylobacter* spp. was the pathogen of greatest prevalence (30.5%; 95% CI: 11.5–59.5%). Observations regarding giblets and minced meat were only found for two pathogens each (*Salmonella* spp. and *S. aureus* in giblets; and *L. monocytogenes* and *Salmonella* spp. in minced meat). *S. aureus* had the highest prevalence in giblets (43.0%; 95% CI: 16.9–73.6%) while *L. monocytogenes* was the pathogen most frequently recovered in minced meat (19.0%; 95% CI: 7.30–40.9%).

This data partition, by type of cut, allows some insight into the sources of heterogeneity between studies. Overall, for *Salmonella* spp., *S. aureus* and *Campylobacter* spp., the different types of meat cuts were able to explain 15 to 25% of this variability (R^2). However, for *L. monocytogenes*, this moderator, cut type, is seemingly not a source of between-study variability.

3.4. Incidence of Pathogens by Sampling Stage

The meta-analysis results on the incidence of pathogens by sampling stage are shown in Table 2. *S. aureus* presented the highest prevalence in poultry meat surveyed at retail level (51.6%; 95% CI: 31.8–70.9%) and at the end-processing stage (38.1%; 95% CI: 14.2–69.7%), followed by *Campylobacter* spp. (44.3%; 95% CI: 28.1–61.8% at retail level; 30.7%; 95% CL: 11.8–59.4% at end-processing). *Salmonella* spp. had the lowest prevalence both at the end-processing stage (5.40%; 95% CI: 1.60–16.1%) and at retail level (10.4%; 95% CI: 5.30–19.3%). Except for *L. monocytogenes*, the recovery of pathogens in poultry meat at retail (i.e., sampled from supermarkets and butcher's) is more frequent than when sampled in factories at the end of processing (Table 2).

Table 2. Meta-analysis of the incidence of pathogens in poultry meat surveyed in Europe by sampling stage.

Pathogen	*n*	Pooled Prevalence	95% CI Pooled Prevalence
Campylobacter spp.			
End-processing	9	0.307	(0.118–0.594)
Retail	41	0.443	(0.281–0.618)

Table 2. *Cont.*

Pathogen	n	Pooled Prevalence	95% CI Pooled Prevalence
L. monocytogenes			
End-processing	14	0.217	(0.084–0.458)
Retail	61	0.171	(0.111–0.255)
Salmonella spp.			
End-processing	16	0.054	(0.016–0.161)
Retail	35	0.104	(0.053–0.193)
S. aureus			
End-processing	5	0.381	(0.142–0.697)
Retail	22	0.516	(0.318–0.709)

3.5. Incidence of Pathogens by Packaging Status

The results of the meta-analysis performed on the incidence of pathogens by packaging status are presented in Table 3. Overall, *Campylobacter* spp. was the pathogen of greatest prevalence in either packed or unpacked poultry meats, with the highest prevalence occurring in chicken (47.2%; 95% CI: 19.4–76.9% in packed chicken; 47.1%; 95% CI: 13.1–84.1% in unpacked chicken). Further, it can be observed that, for the other three bacteria, all unpacked products revealed higher prevalence of pathogens than the packed ones, which can be explained by the fact that packed products have a physical barrier which helps to decrease and slow down the meat deterioration processes [92].

Contrarily to the moderator "type of cut", this moderator "packaging status" explained only marginally the between-study variability present in the data.

Table 3. Meta-analysis of the incidence of pathogens in chicken and other poultry meat by packaging status. Heterogeneity analysis comprises between-study variability to total variability ratio (I^2) and proportion of between-study variability explained by packaging status (R^2).

Microorganism	Product (n)	Pooled Prevalence	95% CI Pooled Prevalence	Heterogeneity
Campylobacter spp.	**Chicken**			
	Packed (20)	0.472	(0.194–0.769)	$I^2 = 0.419$
	Unpacked (9)	0.471	(0.131–0.841)	$R^2 = 0.018$
	Other Poultry			$I^2 = 0.211$
	Packed (18)	0.311	(0.195–0.456)	R^2 = NA *
L. monocytogenes	**Chicken**			
	Packed (37)	0.185	(0.114–0.287)	$I^2 = 0.330$
	Unpacked (8)	0.307	(0.114–0.605)	$R^2 = 0.034$
	Other Poultry			
	Packed (18)	0.125	(0.074–0.201)	$I^2 = 0.408$
	Unpacked (12)	0.148	(0.079–0.260)	$R^2 = 0.047$
Salmonella spp.	**Chicken**			
	Packed (17)	0.031	(0.010–0.097)	$I^2 = 0.455$
	Unpacked (9)	0.048	(0.021–0.108)	$R^2 = 0.007$
	Other Poultry			$I^2 = 0.258$
	Packed (25)	0.079	(0.039–0.150)	R^2 = NA
S. aureus	**Chicken**			$I^2 = 0.483$
	Packed (15)	0.408	(0.165–0.708)	R^2 = NA
	Other Poultry			
	Packed (6)	0.298	(0.152–0.502)	$I^2 = 0.313$
	Unpacked (5)	0.413	(0.189–0.679)	$R^2 = 0.157$

(*) Not applicable.

3.6. Incidence of Pathogens by Cold Preservation Type

The results of the meta-analysis on the incidence of pathogens by cold preservation type are shown in Table 4. In chilled poultry meat, *S. aureus* was the pathogen of highest prevalence (46.9%; 95%

CI: 30.8–63.7%), followed by *Campylobacter* spp. (43.9%; 95% CI: 24.2–65.7%). *Salmonella* spp. shows the lowest prevalence (7.10%; 95% CI: 4.10–12.0%) in chilled meat.

As few data available were available for freezing preservation, conclusions regarding contaminated frozen poultry meat can only be drawn for one pathogen, *Campylobacter* spp. From the prevalence values obtained in chilled and frozen meats, it can be stated that freezing affects the growth and recovery of *Campylobacter* spp., thus ensuring lower prevalence of this pathogen in poultry meats (9.80%; 95% CI: 3.20–26.3%) and improving food safety.

To gather the incidence measures retrieved from primary studies for *Campylobacter* spp. and *Salmonella* spp. in chicken meat, forest plots were built (Figures 2 and 3). These pathogens were selected to assess their incidence per European country as they are both great concerns in terms of food safety, but one has currently very high incidence in this type of product, with more control still being needed, while the other is nowadays better controlled, after the implementation of various control processes specific to *Salmonella* spp.

Figure 2 reveals that the highest overall frequencies of *Campylobacter* spp. in chicken meat were reported in Italy (99.5%; 95% CI: 96.7–99.9%) and Northern Ireland (93.7%; 95% CI: 84.3–97.6%). Regarding *Salmonella* spp., Figure 3 shows that Turkey and Spain reported the highest overall frequencies of this pathogen in chicken meat, with prevalence values as high as 58.3% (95% CI: 38.3–75.9%) and 33.3% (95% CI: 4.30–84.6%), respectively.

Conclusions should be carefully taken from these plots, as the results might reflect more than just values of prevalence: for example, in Spain, coverage of the surveillance system for campylobacteriosis has improved and the number of reported confirmed cases has almost doubled since 2012 [91]. In any case, it can be stated that these forest plots consolidate the fact that *Salmonella* spp. is currently better controlled (from the incidence point of view but also the quality of the surveillance system) throughout the entire food chain in comparison to *Campylobacter* spp., as the prevalence of the latter is significantly higher than the first. It also emphasizes the pressing need of establishing control processes for *Campylobacter* spp. as effective as the ones in place for *Salmonella* spp., and that might not be the same in different countries, as previously stated by Skarp et al. (2015) [93].

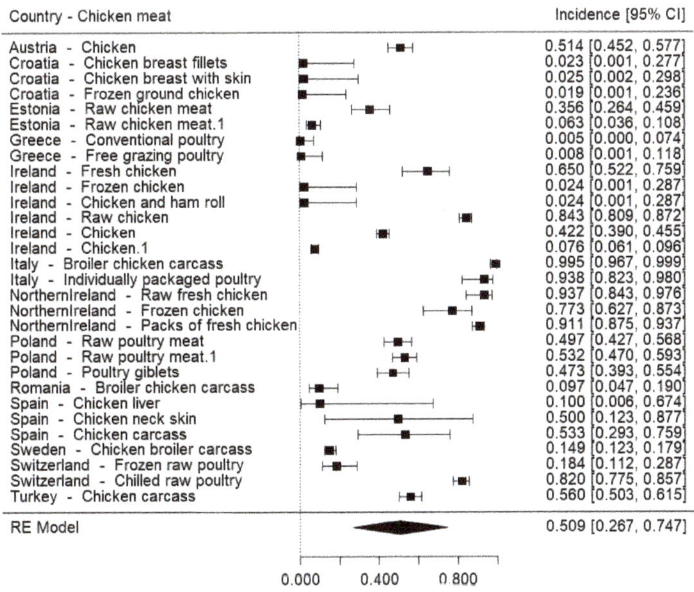

Figure 2. Forest plot of the incidence of *Campylobacter* spp. in chicken meat surveyed in European establishments.

Table 4. Meta-analysis of the incidence of pathogens in poultry meat by cold preservation type.

Stage	Microorganism	*n*	Pooled Prevalence	95% CI Pooled Prevalence
Chilled	*Campylobacter* spp.	45	0.439	(0.242–0.657)
	L. monocytogenes	74	0.177	(0.126–0.243)
	Salmonella spp.	48	0.071	(0.041–0.121)
	S. aureus	26	0.469	(0.308–0.637)
Frozen	*Campylobacter* spp.	5	0.098	(0.032–0.263)

Country - Chicken meat	Incidence [95% CI]
Austria - Chicken	0.164 [0.125, 0.212]
Croatia - Chicken breast fillets	0.143 [0.047, 0.361]
Croatia - Chicken breast with skin	0.105 [0.026, 0.337]
Croatia - Frozen ground chicken	0.115 [0.038, 0.303]
Estonia - Raw broiler chicken meat	0.040 [0.019, 0.082]
Greece - Conventional poultry	0.005 [0.000, 0.074]
Ireland - Raw chicken meat	0.025 [0.023, 0.027]
Ireland - Raw chicken	0.051 [0.035, 0.074]
Ireland - Fresh chicken	0.015 [0.005, 0.044]
Romania - Broiler chicken carcass	0.042 [0.014, 0.121]
Spain - Chicken whole carcass	0.083 [0.005, 0.622]
Spain - Chicken breast and leg skin	0.333 [0.043, 0.846]
Spain - Chicken neck skin	0.125 [0.007, 0.734]
Spain - Chicken liver	0.125 [0.007, 0.734]
Spain - Fresh chicken leg/breast	0.143 [0.020, 0.581]
Spain - Frozen chicken leg/wing	0.167 [0.023, 0.631]
Sweden - Broilers meat	0.023 [0.007, 0.068]
Sweden - Chicken broiler carcass	0.001 [0.000, 0.013]
Turkey - Chicken parts	0.006 [0.001, 0.041]
Turkey - Chicken meat parts	0.583 [0.383, 0.759]
Turkey - Chicken carcass	0.160 [0.061, 0.357]
Turkey - Chicken legs	0.080 [0.020, 0.269]
Turkey - Chicken wings	0.120 [0.039, 0.313]
Turkey - Chicken breast meat	0.480 [0.296, 0.669]
Turkey - Chicken giblets	0.080 [0.020, 0.269]
UK - Raw wings chicken	0.008 [0.002, 0.030]
RE Model	0.045 [0.020, 0.097]

0.000 0.400 0.800

Figure 3. Forest plot of the incidence of *Salmonella* spp. in chicken meat surveyed in European establishments.

4. Discussion

4.1. Incidence of *S. aureus*

The general meta-analysis, that is, with all data collected without any partitions, reveals that *S. aureus* is the pathogen of highest prevalence in overall poultry meats. According to Pepe et al. (2006) [94], staphylococci is one of the most common bacteria present in poultry slaughtering and processing environments, which is corroborated by the results of the sampling stage meta-analysis, in which *S. aureus* presented the highest prevalence in chicken meat surveyed at the end of processing (38.1%; 95% CI: 14.2–69.7%), but also at retail level (51.6%; 95% CI: 31.8–70.9%). The high prevalence at the end-processing stage is likely to be a consequence of the lack of control steps or the inefficiency to avoid contamination by this pathogen from the beginning of the process, at the slaughterhouses, where the skin and mucous membranes of animals are often contaminated, whether it is naturally or due to cross-contamination with infected carcasses. At retail level, there is not always a strict temperature control, making the products susceptible to this pathogen through various routes of contamination. As the vegetative form of *S. aureus* requires temperatures above those used for refrigeration to grow to levels of concentration of public health relevance [8], this is a relatively easy microorganism to control if the cold chain is secured throughout production, in comparison to *L. monocytogenes*, for instance, which grows even at refrigeration temperatures. In this sense, it is easier to reduce the health risks associated with this pathogen, which can explain the currently reduced effects on public health and low number of diseased.

Regarding the different cutting types, whole carcasses revealed to be the product with highest prevalence of pathogens, in particularly *S. aureus*, the one of greatest incidence. The trend observed among all analyzed microorganisms, that greater incidence occurs in whole carcasses in comparison with cut/minced meat and giblets, may be associated with the microbiological detection method used. In whole carcasses, microbiological analysis is preceded by homogenization of the entire carcass [54], meaning that the entire surface is sampled. On the contrary, analysis of pre-cut or minced meat and giblets are carried out by sampling 25 g of the product or swabbing 100 cm^2 of the surface [75]. As it would be expected, sampling of whole carcasses is a much more sensitive technique in terms of pathogen detection in comparison to smaller samples of 25 g or 100 cm^2 swabs, hence the higher prevalence detected.

For a mesophilic bacterium, *S. aureus* has a relatively high heat resistance [95], which can explain why some of the control steps implemented are not suitable for its inactivation. In particular, packaging was found insufficient to reduce the presence of this pathogen, as packed poultry meats revealed considerable prevalence of *S. aureus*, which may be caused by inefficient sterilization of the packaging material. Besides having a physical barrier to microorganisms, packed products are also subjected to a specific gas composition. Although the gas composition certainly impacts the spoilage of poultry meats, it is necessary to optimize the gas mixture that will affect bacterial growth the most [96].

Overall, special attention must be paid to *S. aureus* in undercooked meats and in chicken that will be used to produce transformed products such as sausages, as it is possible for *S. aureus* to survive and proliferate during further meat processing.

4.2. Incidence of *Campylobacter* spp.

The general meta-analysis showed that *Campylobacter* spp. is a pathogen of high prevalence in European poultry meats, which agrees with the statements of Humphrey et al. (2007), where poultry was pointed as an important source of campylobacteriosis, mostly because this bacterium is often carried in the intestinal tract of such animals [97].

Campylobacter spp. has a great colonization capacity, higher levels of intestinal carriage at slaughter than other bacterium (for instance *Salmonella* spp.) [98], and it is highly resistant to procedures such as scalding, washing and cooling, which generally reduce the incidence of other microorganisms to acceptable levels [97]. The reason for this survival may be the unique ability to attach to poultry tissues during carcass processing [98]. The incidence of *Campylobacter* spp. in the many chilled poultry meat samples surveyed across Europe confirms that, while cooling may be able to reduce the fast proliferation that generally occurs during the slaughter process, yet, it is insufficient to inactivate this pathogen [98,99]. The significantly lower prevalence estimated for frozen poultry (~10%), as opposed to the ~44% pooled prevalence in chilled poultry, was expected since freezing is known to decrease the number of campylobacters in time. In a fate study [100], it was found that *C. jejuni* inoculated in poultry meat samples decreased from 7.5 log CFU/g to 3.8 log CFU/g after only 30 min storage at −20 °C.

Regardless of the type of cut or packaging status, in chicken and other poultry meats, high incidence of this pathogen was observed. Furthermore, this meta-analysis also revealed high prevalence of *Campylobacter* spp. at the end of processing and retail level. Recently, studies are showing that campylobacters may be more robust than previously thought and represent a superior challenge to food safety [95], reason as to why the implemented safety control procedures during meat processing might be scarce or inadequate. With farms being the preliminary site of *Campylobacter* entry into production, the major intervention strategies should be targeted at farm level, enhancing biosecurity and implementing better monitoring, as interventions at slaughter process are less efficient [93].

There are plenty of opportunities for improvement along the food chain when it comes to the overall goal of reducing campylobacteriosis, with new and effective measures needing to be quickly implemented as *Campylobacter* has been the most commonly reported gastrointestinal bacterial pathogen in humans in the EU since 2005 [91]. In particular, *C. jejuni* and *C. coli* have been reported as the main causes of human campylobacteriosis [91,93].

4.3. Incidence of L. monocytogenes

Considered by EFSA as the second leading cause of food poisoning outbreaks [91], *Listeria monocytogenes* was found to be, in this meta-analysis, the third most incident pathogen in chicken and other poultry meats.

L. monocytogenes can be found naturally in the environment (soil, sewage, animal feces and water) [99,101] and in foods. In particular, poultry can be asymptomatic carriers of this pathogen and introduce contamination in slaughterhouses [102,103]. This is a great concern at the industrial level because this pathogen has the ability to withstand and adapt to various environmental stresses: it can multiply at low temperatures and form biofilms on food-processing equipment and food-contact surfaces, causing cross-contamination in chicken meat and its derivatives [102] and making a wide range of foodstuffs susceptible to contamination, which is particularly concerning for RTE foods that are not cooked or re-heated. The formation of these biofilms can be the cause for the high incidence values of *L. monocytogenes* found at the end-processing stage and retail level.

Commonly found in bird feces [104], *L. monocytogenes* high incidence in carcasses is likely to arise from cross-contamination during processing. At the end of the process, cooling of meat is intended to inhibit the multiplication of pathogens, but it was observed that in chilled poultry meats, the prevalence of *L. monocytogenes* was still quite relevant, confirming that refrigeration is not enough to inhibit the proliferation of this cold-resistant pathogen.

Following what was already observed for other pathogens, the occurrence of *L. monocytogenes* is higher in broiler meat commercialized without packaging than in packed ones. Packaging can stop further cross-contamination after the products are packed, but if the product is already contaminated with high levels of pathogens, if the wrapper is not properly sterilized, or if the inactivation processes used after packaging are not adequate, presence of pathogens can still occur.

4.4. Incidence of Salmonella spp.

In this work, we did not verify high levels of incidence of *Salmonella* spp. in poultry meat when compared to the other three pathogens under study. The highest prevalence found was in whole carcasses of poultry meat other than chicken, which is probably a consequence of the microbiological detection method, as previously stated.

It is evident that the low incidence of *Salmonella* spp. observed in this work is due to the high investments in zoonoses control in the past years. Since 2003, EU Member States have the responsibility to implement *Salmonella* national control programs and report the results to the European Commission and EFSA, as part of the annual EU zoonoses monitoring. These programs aim to reduce *Salmonella* prevalence in certain animal populations, particularly in breeding flocks of *Gallus gallus*, laying hens, broilers and breeding and fattening turkeys [91].

Currently, the establishment of strict biosecurity measures at farm level (including *Salmonella*-free poultry feed and water), vaccination programs in the parent flocks, and testing/removal of positive flocks from production have been used as control measures [104]. Additionally, the use of feed additives and acidified food and water have been encouraged, as the pH reduction is expected to have a protective effect on the feed, milling and feeding equipment and on the general environment [104].

This large reduction of salmonellosis and *Salmonella* prevalence in food products in the past years should be taken as a motivation for the control of other pathogens, as the effective implementation of control programs demonstrate that it is possible to produce safer food if the proper, adequate measures are implemented. However, it is crucial to keep in mind that the efficacy of such interventions depends on the level of bacterial contamination and that the control steps must be specific for each microorganism, as they have their own characteristics and different resistance to diverse conditions.

Despite the declining trend of salmonellosis in the EU that can give an increased sense of security, continuous research is mandatory to solve new difficulties, such as the increasing antimicrobial resistance in non-typhoid *Salmonella* species that has become a serious problem for public health worldwide [104].

Moreover, aiming for the reduction of one specific serovar or serotype is not sufficient, as the currently predominant serotypes in poultry flocks are likely to change over time [104].

5. Conclusions

It is a great concern that chicken meat has been considered one of the main causes of food poisoning while the poultry industry is simultaneously one of the most important sources of animal protein for the world's population [102]. Pathogenic contamination of poultry can occur at any level: in the initial production environment, through vertical transmission (via egg, triggering the birth of infected chicks) or horizontal transmission (caused by contaminated environment or feed); or in the slaughtering process [6,47]. With gastrointestinal tract of birds and slaughtering facilities identified as the main reservoirs of poultry meat contaminants [96], it is necessary to implement preventive and corrective measures in several stages, but mainly on farm to reduce the initial levels of contamination that enter the process. As every stage of poultry meat production and processing systems has its own unique challenges regarding pathogen contamination and control, a multi-hurdle approach is likely to be the best strategy for pathogen reduction and elimination [105].

This meta-analysis on incidence data from European surveys indicated that *S. aureus* is currently the main contaminating pathogen of poultry meat, followed closely by *Campylobacter* spp. The establishment of control processes specific to these pathogens, as it was done in *Salmonella* spp. control programs, will certainly have a great impact on their current values of incidence, enabling the production and provision of safer meat products to consumers.

Adequate interventions at the processing stages can be assessed through challenge tests and predictive microbiology. In particular, growth and inactivation models can take into account factors such as the levels of contamination when carcasses leave the processing plant, storage time in retail stores, transport time, storage times in homes and the temperatures carcasses were exposed to during each of these periods [104], making this a unique tool for researchers and food companies to increase food safety and prevent new outbreaks.

Despite industries' responsibility to take action on these matters, consumers should be further educated and encouraged to take preventive measures to ensure their health and well-being, such as: sanitization of hands, surfaces and utensils before and after handling poultry meat; separation of raw poultry meat from other foods (especially cooked) to avoid cross-contamination; proper storage of products, ensuring refrigeration temperatures under 5 °C; consumption of properly cooked, non-washed, poultry meat, as washing can spread bacteria and contaminate kitchen surfaces.

Author Contributions: V.R. performed the bibliographic searches and extracted meta-analytical data; A.G.-T. double-checked extracted data, adjusted meta-analysis models and wrote 20% of the article; B.N.S. wrote 60% of the article and revised it; U.G.-B. and V.C. conceived and designed the work, coded data analysis scripts, produced meta-analytical graphs, wrote 20% of the article, revised it and proofread it.

Acknowledgments: Gonzales-Barron wishes to acknowledge the financial support provided by the Portuguese Foundation for Science and Technology (FCT) through the award of a five-year Investigator Fellowship (IF) in the mode of Development Grants (IF/00570).

Conflicts of Interest: The authors declare no conflict of interest.

References

1. Van Nierop, W.; Dusé, A.G.; Marais, E.; Aithma, N.; Thothobolo, N.; Kassel, M.; Stewart, R.; Potgieter, A.; Fernandes, B.; Galpin, J.S.; et al. Contamination of chicken carcasses in Gauteng, South Africa, by *Salmonella*, *Listeria monocytogenes* and *Campylobacter*. *Int. J. Food Microbiol.* **2005**, *99*, 1–6. [CrossRef] [PubMed]
2. Hennekinne, J.-A.; De Buyser, M.-L.; Dragacci, S. Staphylococcus aureus and its food poisoning toxins: Characterization and outbreak investigation. *FEMS Microbiol. Rev.* **2012**, *36*, 815–836. [CrossRef] [PubMed]
3. Centers for Disease Control and Prevention. Food Safety: Symptoms and Sources of Food Poisoning. 2018. Available online: http://www.cdc.gov/foodsafety/symptoms.html (accessed on 23 April 2018).
4. Balaban, N.; Rasooly, A. Staphylococcal enterotoxins. *Int. J. Food Microbiol.* **2000**, *61*, 1–10. [CrossRef]

5. Jørgensen, F.; Bailey, R.; Williams, S.; Henderson, P.; Wareing, D.R.A.; Bolton, F.J.; Ward, L.; Humphrey, T.J. Prevalence and numbers of *Salmonella* and *Campylobacter* spp. on raw, whole chickens in relation to sampling methods. *Int. J. Food Microbiol.* **2002**, *76*, 151–164. [CrossRef]
6. Miettinen, M.; Palmu, L.; Bjorkroth, K.; Korkeala, H. Prevalence of *Listeria monocytogenes* in broilers at the abattoir, processing plant, and retail level. *J. Food Prot.* **2001**, *64*, 994–999. [CrossRef] [PubMed]
7. EFSA Panel on Biological Hazards (BIOHAZ). Scientific Opinion on quantification of the risk posed by broiler meat to human campylobacteriosis in the EU. *EFSA J.* **2010**, *8*, 1437.
8. EFSA Panel on Biological Hazards (BIOHAZ); EFSA Panel on Contaminants in the Food Chain (CONTAM); EFSA Panel on Animal Health and Welfare (AHAW). Scientific Opinion on the public health hazards to be covered by inspection of meat (poultry). *EFSA J.* **2012**, *10*, 2741.
9. Pires, S.; de Knegt, L.; Hald, T. *Estimation of the Relative Contribution of Different Food and Animal Sources to Human Salmonella Infections in the European Union*; Question No EFSA-Q-2010-00685; EFSA Supporting Publications: Parma, Italy, 2011.
10. Gonzales-Barron, U.; Butler, F. The use of meta-analytical tools in risk assessment for food safety. *Food Microbiol.* **2011**, *28*, 823–827. [CrossRef] [PubMed]
11. Xavier, C.; Gonzales-Barron, U.; Paula, V.; Estevinho, L.; Cadavez, V. Meta-analysis of the incidence of foodborne pathogens in Portuguese meats and their products. *Food Res. Int.* **2014**, *55*, 311–323. [CrossRef]
12. Akpolat, N.; Elci, S.; Atmaca, S.; Gül, K. *Listeria monocytogenes* in products of animal origin in Turkey. *Vet. Res. Commun.* **2004**, *28*, 561–567. [CrossRef] [PubMed]
13. Alonso-Hernando, A.; Prieto, M.; García-Fernández, C.; Alonso-Calleja, C.; Capita, R. Increase over time in the prevalence of multiple antibiotic resistance among isolates of *Listeria monocytogenes* from poultry in Spain. *Food Control* **2012**, *23*, 37–41. [CrossRef]
14. Álvarez-Astorga, M.; Capita, R.; Alonso-Calleja, C.; Moreno, B.; García-Fernández, M. Microbiological quality of retail chicken by-products in Spain. *Meat. Sci.* **2002**, *62*, 45–50. [CrossRef]
15. Angelidis, A.; Koutsoumanis, K. Prevalence and concentration of *Listeria monocytogenes* in sliced ready-to-eat meat products in the Hellenic retail market. *J. Food Prot.* **2006**, *69*, 938–942. [CrossRef] [PubMed]
16. Antunes, P.; Reu, C.; Sousa, J.; Pestana, N.; Peixe, L. Incidence and susceptibility to antimicrobial agents of *Listeria* spp. and *Listeria monocytogenes* isolated from poultry carcasses in Porto, Portugal. *J. Food Prot.* **2002**, *65*, 1888–1893. [CrossRef] [PubMed]
17. Aras, Z.; Ardic, M. Occurrence and Antibiotic Susceptibility of *Listeria* Species in Turkey Meats. *Korean J. Food Sci. Anim. Resour.* **2015**, *35*, 669–673. [CrossRef] [PubMed]
18. Ayaz, N.D.; Erol, I. Rapid Detection of *Listeria monocytogenes* in ground turkey by immunomagnetic separation and polymerase chain reaction. *J. Rapid Meth. Aut. Microbiol.* **2009**, *17*, 214–222. [CrossRef]
19. Aydin, F.; Gumussoy, K.S.; Ica, T.; Sumerkan, B.; Esel, D.; Akan, M.; Ozdemir, A. The prevalence of *Campylobacter jejuni* in various sources in Kayseri, Turkey, and molecular analysis of isolated strains by PCR-RFLP. *Turk. J. Vet. Anim. Sci.* **2007**, *31*, 13–19.
20. Beninati, C.; Reich, F.; Muscolino, D.; Giarratana, F.; Panebianco, A.; Klein, G.; Atanassova, V. ESBL-producing bacteria and MRSA isolated from poultry and turkey products imported from Italy. *Czech J. Food Sci.* **2015**, *33*, 97–102. [CrossRef]
21. Boer, E.; Zwartkruis-Nahuis, J.; Wit, B.; Huijsdens, X.; Neeling, A.; Bosch, T.; van Oosterom, R.; Vila, A.; Heuvelink, A. Prevalence of methicillin-resistant *Staphylococcus aureus* in meat. *Int. J. Food Microbiol.* **2009**, *134*, 52–56. [CrossRef] [PubMed]
22. Busani, L.; Cigliano, A.; Taioli, E.; Caligiuri, V.; Chiavacci, L.; Di Bella, C.; Battisti, A.; Duranti, A.; Gianfranceschi, M.; Nardella, M.C.; et al. Prevalence of *Salmonella enterica* and *Listeria monocytogenes* contamination in foods of animal origin in Italy. *J. Food Prot.* **2005**, *68*, 1729–1733. [CrossRef] [PubMed]
23. Capita, R.; Alonso-Calleja, C.; Del Camino García-Fernández, M.; Moreno, B. Microbiological quality of retail poultry carcasses in Spain. *J. Food Prot.* **2001**, *64*, 1961–1966. [CrossRef] [PubMed]
24. Capita, R.; Alonso-Calleja, C.; Moreno, B.; García-Fernández, M. Assessment of Baird-Parker Agar as Screening Test for Determination of *Staphylococcus aureus* in Poultry Meat. *J. Microbiol.* **2001**, *39*, 321–325.
25. Capita, R.; Alonso-Calleja, C.; Moreno, B.; García-Fernández, M. Occurrence of *Listeria* species in retail poultry meat and comparison of a cultural/immunoassay for their detection. *Int. J. Food Microbiol.* **2001**, *65*, 75–82. [CrossRef]

26. Capita, R.; Alonso-Calleja, C.; Prieto, M.; Del Camino García-Fernández, M.; Moreno, B. Comparison of PALCAM and modified oxford plating media for isolation of *Listeria* species in poultry meat following UVM II or fraser secondary enrichment broths. *Food Microbiol.* **2001**, *18*, 555–563. [CrossRef]
27. Carp-Carare, C.; Vlad-Sabie, A.; Floristean, V.-C. Detection and serotyping of *Listeria monocytogenes* in some food products from North-East of Romania. *Rom. Rev. Lab. Med.* **2013**, *21*, 285–292. [CrossRef]
28. Cetinkaya, F.; Cibik, R.; Soyutemiz, G.E.; Ozakin, C.; Kayali, R.; Levent, B. *Shigella* and *Salmonella* contamination in various foodstuffs in Turkey. *Food Control* **2008**, *19*, 1059–1063. [CrossRef]
29. Cetinkaya, F.; Mus, T.E.; Yibar, A.; Guclu, N.; Tavsanli, H.; Cibik, R. Prevalence, serotype identification by multiplex polymerase chain reaction and antimicrobial resistance patterns of *Listeria monocytogenes* isolated from retail foods. *J. Food Saf.* **2014**, *34*, 42–49. [CrossRef]
30. Ceylan, Z.G.; Demirkaya, A.K.; Adiguzel, G. Incidence of *Listeria monocytogenes* in retail chicken meat and establishing relationship with some bacteria by logistic regression. *J. Food Qual.* **2008**, *31*, 121–130. [CrossRef]
31. Cloak, O.; Duffy, G.; Sheridan, J.; Blair, I.; McDowell, D. A survey on the incidence of *Campylobacter* spp. and the development of a surface adhesion polymerase chain reaction (SA-PCR) assay for the detection of *Campylobacter jejuni* in retail meat products. *Food Microbiol.* **2001**, *18*, 287–298. [CrossRef]
32. Cui, Y.; Guran, H.S.; Harrison, M.A.; Hofacre, C.L.; Alali, W.Q. *Salmonella* Levels in Turkey Neck Skins, Drumstick Bones, and Spleens in Relation to Ground Turkey. *J. Food Prot.* **2015**, *78*, 1945–1953. [CrossRef] [PubMed]
33. Dan, S.; Tăbăran, A.; Mihaiu, L.; Mihaiu, M. Antibiotic susceptibility and prevalence of foodborne pathogens in poultry meat in Romania. *J. Infect. Dev. Coun.* **2015**, *9*, 035–041. [CrossRef] [PubMed]
34. Duggan, S.; Jordan, E.; Gutierrez, M.; Barrett, G.; O'Brien, T.; Hand, D.; Kenny, K.; Fanning, J.; Leonard, N.; Egan, J. *Salmonella* in meats, water, fruit and vegetables as disclosed from testing undertaken by Food Business Operators in Ireland from 2005 to 2009. *Irish Vet. J.* **2012**, *65*, 17. [CrossRef] [PubMed]
35. Durul, B.; Acar, S.; Bulut, E.; Kyere, E.O.; Soyer, Y. Subtyping of *Salmonella* Food Isolates Suggests the Geographic Clustering of Serotype Telaviv. *Foodborne Pathog. Dis.* **2015**, *12*, 958–965. [CrossRef] [PubMed]
36. Egervarn, M.; Borjesson, S.; Byfors, S.; Finn, M.; Kaipe, C.; Englund, S.; Lindblad, M. *Escherichia coli* with extended-spectrum beta-lactamases or transferable AmpC beta-lactamases and *Salmonella* on meat imported into Sweden. *Int. J. Food Microbiol.* **2014**, *171*, 8–14. [CrossRef] [PubMed]
37. Elmali, M.; Can, H.Y.; Yaman, H. Prevalence of *Listeria monocytogenes* in poultry meat. *Food Sci. Technol.* **2015**, *35*, 672–675. [CrossRef]
38. Filiousis, G.; Johansson, A.; Frey, J.; Perreten, V. Prevalence, genetic diversity and antimicrobial susceptibility of *Listeria monocytogenes* isolated from open-air food markets in Greece. *Food Control* **2009**, *20*, 314–317. [CrossRef]
39. Fox, E.M.; Wall, P.G.; Fanning, S. Control of *Listeria* species food safety at a poultry food production facility. *Food Microbiol.* **2015**, *51*, 81–86. [CrossRef] [PubMed]
40. Gundogan, N.; Citak, S.; Yucel, N.; Devren, A. A note on the incidence and antibiotic resistance of *Staphylococcus aureus* isolated from meat and chicken samples. *Meat Sci.* **2005**, *69*, 807–810. [CrossRef] [PubMed]
41. Guven, K.; Mutlu, M.; Gulbandilar, A.; Cakir, P. Occurrence and characterization of *Staphylococcus aureus* isolated from meat and dairy products consumed in Turkey. *J. Food Saf.* **2010**, *30*, 196–212. [CrossRef]
42. Iseri, O.; Erol, I. Incidence and antibiotic resistance of *Salmonella* spp. in ground turkey meat. *Br. Poult. Sci.* **2010**, *51*, 60–66. [CrossRef] [PubMed]
43. Kalender, H. Prevalence of *Listeria* Species in Ground Beef and Chicken Meat Sold in Eastern Turkey. *Pak. Vet. J.* **2012**, *32*, 456–458.
44. Kilic, S.; Kuplulu, O. Detection the enterotoxin producing capacity of coagulase positive *Staphylococcus* by EIA (Enzyme Immuno Assay) isolated from turkey meat. *Ankara Univ. Vet. Fak. Derg.* **2009**, *56*, 183–186.
45. Koluman, A.; Unlu, T.; Dikici, A.; Tezel, A.; Akcelik, E.N.; Burkan, Z.T. Presence of *Staphylococcus aureus* and Staphylococcal Enterotoxins in Different Foods. *Kafkas Univ. Vet. Fak. Derg.* **2011**, *17*, S55–S60.
46. Kosek-Paszkowska, K.; Bania, J.; Bystron, J.; Molenda, J.; Czerw, M. Occurrence of *Listeria* sp. in raw poultry meat and poultry meat products. *Bull. Vet. Inst. Pulawy* **2005**, *49*, 219–222.
47. Kozačinski, L.; Fleck, Z.; Kozačinski, Z.; Filipovic, I.; Mitak, M.; Bratulic, M.; Mikuš, T. Evaluation of shelf life of pre-packed cut poultry meat. *Vet. Arh.* **2012**, *82*, 47–58.
48. Kozačinski, L.; Hadžiosmanović, M.; Zdolec, N. Microbiological quality of poultry meat on the Croatian market. *Vet. Arh.* **2006**, *76*, 305–313.

49. Kramarenko, T.; Nurmoja, I.; Kaerssin, A.; Meremaee, K.; Horman, A.; Roasto, M. The prevalence and serovar diversity of *Salmonella* in various food products in Estonia. *Food Control* **2014**, *42*, 43–47. [CrossRef]
50. Kramarenko, T.; Roasto, M.; Meremaee, K.; Kuningas, M.; Poltsama, P.; Elias, T. *Listeria monocytogenes* prevalence and serotype diversity in various foods. *Food Control* **2013**, *30*, 24–29. [CrossRef]
51. Lambertz, S.T.; Nilsson, C.; Bradenmark, A.; Sylven, S.; Johansson, A.; Jansson, L.-M.; Lindblad, M. Prevalence and level of *Listeria monocytogenes* in ready-to-eat foods in Sweden 2010. *Int. J. Food Microbiol.* **2012**, *160*, 24–31. [CrossRef] [PubMed]
52. Latorre, L.; Parisi, A.; Fraccalvieri, R.; Normanno, G.; La Porta, M.C.N.; Goffredo, E.; Palazzo, L.; Ciccarese, G.; Addante, N.; Santagada, G. Low prevalence of *Listeria monocytogenes* in foods from Italy. *J. Food Prot.* **2007**, *70*, 1507–1512. [CrossRef] [PubMed]
53. Ledergerber, U.; Regula, G.; Stephan, R.; Danuser, J.; Bissig, B.; Stark, K. Risk factors for antibiotic resistance in *Campylobacter* spp. isolated from raw poultry meat in Switzerland. *BMC Public Health* **2003**, *3*, 39. [CrossRef] [PubMed]
54. Lindblad, M.; Lindmark, H.; Lambertz, S.T.; Lindqvist, R. Microbiological baseline study of broiler chickens at Swedish slaughterhouses. *J. Food Prot.* **2006**, *69*, 2875–2882. [CrossRef] [PubMed]
55. Little, C.L.; Sagoo, S.K.; Gillespie, I.A.; Grant, K.; McLauchlin, J. Prevalence and Level of *Listeria monocytogenes* and Other *Listeria* Species in Selected Retail Ready-to-Eat Foods in the United Kingdom. *J. Food Prot.* **2009**, *72*, 1869–1877. [CrossRef] [PubMed]
56. Mackiw, E.; Rzewuska, K.; Stos, K.; Jarosz, M.; Korsak, D. Occurrence of *Campylobacter* spp. in Poultry and Poultry Products for Sale on the Polish Retail Market. *J. Food Prot.* **2011**, *74*, 986–989. [CrossRef] [PubMed]
57. Madden, R.H.; Moran, L.; Scates, P.; McBride, J.; Kelly, C. Prevalence of *Campylobacter* and *Salmonella* in Raw Chicken on Retail Sale in the Republic of Ireland. *J. Food Prot.* **2011**, *74*, 1912–1916. [CrossRef] [PubMed]
58. Malpass, M.; Williams, A.; Jones, D.; Omed, H. Microbiological quality of chicken wings damaged on the farm or in the processing plant. *Food Microbiol.* **2010**, *27*, 521–525. [CrossRef] [PubMed]
59. Manfreda, G.; De Cesare, A.; Bondioli, V.; Stern, N.; Franchini, A. Enumeration and identity of *Campylobacter* spp. in Italian broilers. *Poult. Sci.* **2006**, *85*, 556–562. [CrossRef] [PubMed]
60. Manios, S.G.; Grivokostopoulos, N.C.; Bikouli, V.C.; Doultsos, D.A.; Zilelidou, E.A.; Gialitaki, M.A.; Skandamis, P.N. A 3-year hygiene and safety monitoring of a meat processing plant which uses raw materials of global origin. *Int. J. Food Microbiol.* **2015**, *209*, 60–69. [CrossRef] [PubMed]
61. Mayrhofer, S.; Paulsen, P.; Smulders, F.; Hilbert, F. Antimicrobial resistance profile of five major food-borne pathogens isolated from beef, pork and poultry. *Int. J. Food Microbiol.* **2004**, *97*, 23–29. [CrossRef] [PubMed]
62. Meldrum, R.J.; Charles, D.; Mannion, P.; Ellis, P. Variation in the microbiological quality of commercially produced vacuum-packed cooked sliced meat between production and the end of shelf-life. *Int. J. Environ. Res.* **2014**, *24*, 269–277. [CrossRef] [PubMed]
63. Mena, C.; Almeida, G.; Carneiro, L.; Teixeira, P.; Hogg, T.; Gibbs, P. Incidence of *Listeria monocytogenes* in different food products commercialized in Portugal. *Food Microbiol.* **2004**, *21*, 213–216. [CrossRef]
64. Miranda, J.N.; Vazquez, B.I.; Fente, C.A.; Calo-Mata, P.; Cepeda, A.; Franco, C.M. Comparison of Antimicrobial Resistance in *Escherichia coli, Staphylococcus aureus*, and *Listeria monocytogenes* Strains Isolated from Organic and Conventional Poultry Meat. *J. Food Prot.* **2008**, *71*, 2537–2542. [CrossRef] [PubMed]
65. Moore, J.; Wilson, T.; Wareing, D.; Humphrey, T.; Murphy, P. Prevalence of thermophilic *Campylobacter* spp. in ready-to-eat foods and raw poultry in Northern Ireland. *J. Food Prot.* **2002**, *65*, 1326–1328. [CrossRef] [PubMed]
66. Moran, L.; Scates, P.; Madden, R.H. Prevalence of *Campylobacter* spp. in Raw Retail Poultry on Sale in Northern Ireland. *J. Food Prot.* **2009**, *72*, 1830–1835. [CrossRef] [PubMed]
67. Navas, J.; Ortiz, S.; López, P.; López, V.; Martínez-Suárez, J. Different enrichment procedures for recovery of *Listeria monocytogenes* from raw chicken samples can affect the results of detection (by chromogenic plating or real-time PCR) and lineage or strain identification. *J. Food Prot.* **2007**, *70*, 2851–2854. [CrossRef] [PubMed]
68. Ojeniyi, B.; Christensen, J.; Bisgaard, M. Comparative investigations of *Listeria monocytogenes* isolated from a turkey processing plant, turkey products, and from human cases of listeriosis in Denmark. *Epidemiol. Infect.* **2000**, *125*, 303–308. [CrossRef] [PubMed]
69. Ormanci, F.S.B.; Erol, I.; Ayaz, N.D.; Iseri, O.; Sariguzel, D. Immunomagnetic separation and PCR detection of *Listeria monocytogenes* in turkey meat and antibiotic resistance of the isolates. *Br. Poult. Sci.* **2008**, *49*, 560–565. [CrossRef] [PubMed]

70. Özdemir, H.; Keyvan, E. Isolation and characterisation of *Staphylococcus aureus* strains isolated from beef, sheep and chicken meat. *Ankara Univ Vet. Fak. Derg.* **2016**, *63*, 333–338.
71. Paludi, D.; Vergara, A.; Festino, A.; Di Ciccio, P.; Costanzo, C.; Conter, M.; Zanardi, E.; Ghidini, S.; Ianieri, A. Antimicrobial resistance pattern of methicillin-resistant *Staphylococcus aureus* in the food industry. *J. Biol. Regul. Homeost. Agents* **2011**, *25*, 671–677. [PubMed]
72. Praakle-Amin, K.; Hanninen, M.; Korkeala, H. Prevalence and genetic characterization of *Listeria monocytogenes* in retail broiler meat in Estonia. *J. Food Prot.* **2006**, *69*, 436–440. [CrossRef] [PubMed]
73. Rantsiou, K.; Lamberti, C.; Cocolin, L. Survey of *Campylobacter jejuni* in retail chicken meat products by application of a quantitative PCR protocol. *Int. J. Food Microbiol.* **2010**, *141*, S75–S79. [CrossRef] [PubMed]
74. Reiter, M.G.; Lopez, C.; Jordano, R.; Medina, L.M. Comparative Study of Alternative Methods for Food Safety Control in Poultry Slaughterhouses. *Food Anal. Method* **2010**, *3*, 253–260. [CrossRef]
75. Roasto, M.; Praakle, K.; Korkeala, H.; Elias, P.; Hanninen, M. Prevalence of *Campylobacter* in raw chicken meat of Estonian origin. *Arch. Lebensmittelhyg.* **2005**, *56*, 61–62.
76. Siriken, B.; Ayaz, N.D.; Erol, I. *Listeria monocytogenes* in retailed raw chicken meat in Turkey. *Berl. Munch. Tierarztl. Wochenschr.* **2014**, *127*, 43–49. [PubMed]
77. Soultos, N.; Koidis, P.; Madden, R. Presence of *Listeria* and *Salmonella* spp. in retail chicken in Northern Ireland. *Lett. Appl. Microbiol.* **2003**, *37*, 421–423. [CrossRef] [PubMed]
78. Stella, S.; Soncini, G.; Ziino, G.; Panebianco, A.; Pedonese, F.; Nuvoloni, R.; Di Giannatale, E.; Colavita, G.; Alberghini, L.; Giaccone, V. Prevalence and quantification of thermophilic *Campylobacter* spp. in Italian retail poultry meat: Analysis of influencing factors. *Food Microbiol.* **2017**, *62*, 232–238. [CrossRef] [PubMed]
79. Stoyanchev, T.; Vashin, I.; Ring, C.; Atanassova, V. Prevalence of *Campylobacter* spp. in poultry and poultry products for sale on the Bulgarian retail market. *Anton Leeuwenhoek* **2007**, *92*, 285–288. [CrossRef] [PubMed]
80. Trajković-Pavlović, L.; Popović, M.; Novaković, B.; Gusman-Pasterko, V.; Jevtić, M.; Mirilov, J. Occurrence of *Campylobacter, Salmonella, Yersinia enterocolitica* and *Listeria monocytogenes* in some retail food products in Novi Sad. *Cent. Eur. J. Public Health* **2007**, *15*, 167–171. [PubMed]
81. Trajković-Pavlović, L.; Novaković, B.; Martinov-Cvejin, M.; Gusman, V.; Bijelović, S.; Dragnić, N.; Balać, D. How a routine checking of *Escherichia coli* in retailed food of animal origin can protect consumers against exposition to *Campylobacter* spp. and *Listeria monocytogenes*? *Vojnosanit. Pregl.* **2010**, *67*, 627–633. [CrossRef] [PubMed]
82. Vitas, A.; Aguado, V.; Garcia-Jalon, I. Occurrence of *Listeria monocytogenes* in fresh and processed foods in Navarra (Spain). *Int. J. Food Microbiol.* **2004**, *90*, 349–356. [CrossRef]
83. Voidarou, C.; Vassos, D.; Rozos, G.; Alexopoulos, A.; Plessas, S.; Tsinas, A.; Skoufou, M.; Stavropoulou, E.; Bezirtzoglou, E. Microbial challenges of poultry meat production. *Anaerobe* **2011**, *17*, 341–343. [CrossRef] [PubMed]
84. Vossenkuhl, B.; Brandt, J.; Fetsch, A.; Käsbohrer, A.; Kraushaar, B.; Alt, K.; Tenhagen, B.-A. Comparison of spa Types, SCCmec types and antimicrobial resistance profiles of MRSA isolated from Turkeys at farm, slaughter and from retail meat indicates transmission along the production chain. *PLoS ONE* **2014**, *9*, e96308. [CrossRef] [PubMed]
85. Vural, A.; Erkan, M.; Yeşilmen, S. Microbiological quality of retail chicken carcasses and their products in Turkey. *Med. Weter* **2006**, *62*, 1371–1374.
86. Whyte, P.; McGill, K.; Cowley, D.; Madden, R.; Moran, L.; Scates, P.; Carroll, C.; O'Leary, A.; Fanning, S.; Collins, J.; et al. Occurrence of *Campylobacter* in retail foods in Ireland. *Int. J. Food Microbiol.* **2004**, *95*, 111–118. [CrossRef] [PubMed]
87. Yucel, N.; Citak, S.; Onder, M. Prevalence and antibiotic resistance of *Listeria* species in meat products in Ankara, Turkey. *Food Microbiol.* **2005**, *22*, 241–245. [CrossRef]
88. Zadernowska, A.; Chajecka-Wierzchowska, W. Prevalence, biofilm formation and virulence markers of *Salmonella* sp and *Yersinia enterocolitica* in food of animal origin in Poland. *LWT-Food Sci. Technol.* **2017**, *75*, 552–556. [CrossRef]
89. Viechtbauer, W. Conducting Meta-Analyses in R with the metaphor Package. *J. Stat. Softw.* **2010**, *36*, 1–48. [CrossRef]
90. Grothendieck, G. Sqldf: Manipulate R Data Frames Using SQL. R package version 0.4-11. 2017. Available online: https://CRAN.R-project.org/package=sqldf (accessed on 6 April 2018).
91. EFSA. The European Union summary report on trends and sources of zoonoses, zoonotic agents and food-borne outbreaks in 2016. *EFSA J.* **2017**, *15*, e05077.

92. Narasimha Rao, D.; Sachindra, N.M. Modified atmosphere and vacuum packaging of meat and poultry products. *Food Rev. Int.* **2002**, *18*, 263–293. [CrossRef]
93. Skarp, C.P.A.; Hänninen, M.L.; Rautelin, H.I.K. Campylobacteriosis: The role of poultry meat. *Clin. Microbiol. Infect.* **2016**, *22*, 103–109. [CrossRef] [PubMed]
94. Pepe, O.; Blaiotta, G.; Bucci, F.; Anastasio, M.; Aponte, M.; Villani, F. *Staphylococcus aureus* and staphylococcal enterotoxin A in breaded chicken products: Detection and behavior during the cooking process. *Appl. Environ. Microbiol.* **2006**, *72*, 7057–7062. [CrossRef] [PubMed]
95. Stewart, C.M. Chapter 12: *Staphylococcus aureus* and staphylococcal enterotoxins. In *Foodborne Microorganisms of Public Health Significance*, 6th ed.; Hocking, A.D., Ed.; Australian Institute of Food Science and Technology (NSW Branch): Sydney, Australia, 2003; pp. 359–380.
96. Rouger, A.; Tresse, O.; Zagorec, M. Bacterial Contaminants of Poultry Meat: Sources, Species, and Dynamics. *Microorganisms* **2017**, *5*, 50. [CrossRef] [PubMed]
97. Humphrey, T.; O'Brien, S.; Madsen, M. Campylobacters as zoonotic pathogens: A food production perspective. *Int. J. Food Microbiol.* **2007**, *117*, 237–257. [CrossRef] [PubMed]
98. Mead, G.C. Microbiological quality of poultry meat: A review. *Rev. Bras. Cienc. Avic.* **2004**, *6*, 135–142. [CrossRef]
99. Goh, S.G.; Kuan, C.H.; Loo, Y.Y.; Chang, W.S.; Lye, Y.L.; Soopna, P.; Tang, J.Y.; Nakaguchi, Y.; Nishibuchi, M.; Afsah-Hejri, L.; et al. *Listeria monocytogenes* in retailed raw chicken meat in Malaysia. *Poult. Sci.* **2012**, *91*, 2686–2690. [CrossRef] [PubMed]
100. Ivic-Kolevska, S.; Miljkovic-Selimovic, B.; Kocic, B. Survival of *Campylobacter jejuni* in chicken meat at frozen storage temperatures. *Acta Microbiol. Immun. Hung.* **2012**, *59*, 185–198. [CrossRef] [PubMed]
101. Milillo, S.; Stout, J.; Hanning, I.; Clement, A.; Fortes, E.; Den Bakker, H.; Wiedmann, M.; Ricke, S.C. *Listeria monocytogenes* and hemolytic *Listeria innocua* in poultry. *Poult. Sci.* **2012**, *91*, 2158–2163. [CrossRef] [PubMed]
102. Moura, G.F.; Sigarini, C.O.; Figueiredo, E.E.S. *Listeria monocytogenes* in Chicken Meat. *J. Food Nutr. Res.* **2016**, *4*, 436–441.
103. Mbata, T.I. Poultry meat pathogens and its Control. *Int. J. Food Saf.* **2017**, *7*, 20–28.
104. Mouttotou, N.; Ahmad, S.; Kamran, Z.; Koutoulis, K.C. Chapter 12: Prevalence, Risks and Antibiotic Resistance of *Salmonella* in Poultry Production Chain. In *Current Topics in Salmonella and Salmonellosis*; Mares, M., Ed.; InTech: Rijeka, Croatia, 2017; pp. 215–234.
105. Hiett, K.L. Tracing pathogens in poultry and egg production at the abattoir. In *Tracing Pathogens in the Food Chain*; McMeekin, T.A., Ed.; Woodhead Publishing Limited: Cambridge, UK, 2011; pp. 465–502.

Review

Carbon Monoxide in Meat and Fish Packaging: Advantages and Limits

Djamel Djenane [1,*] and Pedro Roncalés [2]

1 Laboratory of Food Quality and Food Safety, Department of Food Science and Technology, University Mouloud Mammeri, P.O. Box 17, Tizi-Ouzou 15000, Algeria

2 Laboratory of Meat and Fish Technology, Department of Animal Production and Food Science, Faculty of Veterinary Sciences, University of Zaragoza, C/Miguel Servet, 177, 50013 Zaragoza, Spain; roncales@unizar.es

* Correspondence: djenane6@yahoo.es or d.djenane@hotmail.com; Tel.: +213-779-001-384; Fax: +213-2621-6819

Received: 23 November 2017; Accepted: 15 January 2018; Published: 23 January 2018

Abstract: Due to increased demands for greater expectation in relation to quality, convenience, safety and extended shelf-life, combined with growing demand from retailers for cost-effective extensions of fresh muscle foods' shelf-life, the food packaging industry quickly developed to meet these expectations. During the last few decades, modified atmosphere packaging (MAP) of foods has been a promising area of research, but much remains to be known regarding the use of unconventional gases such carbon monoxide (CO). The use of CO for meat and seafood packaging is not allowed in most countries due to the potential toxic effect, and its use is controversial in some countries. The commercial application of CO in food packaging was not then considered feasible because of possible environmental hazards for workers. CO has previously been reported to mask muscle foods' spoilage, and this was the primary concern raised for the prohibition, as this may mislead consumers. This review was undertaken to present the most comprehensive and current overview of the widely-available, scattered information about the use of CO in the preservation of muscle foods. The advantages of CO and its industrial limits are presented and discussed. The most recent literature on the consumer safety issues related to the use of CO and consumer acceptance of CO especially in meat packaging systems were also discussed. Recommendations and future prospects were addressed for food industries, consumers and regulators on what would be a "best practice" in the use of CO in food packaging. All this promotes high ethical standards in commercial communications by means of effective regulation, for the benefit of consumers and businesses in the world, and this implies that industrialized countries and members of their regulatory agencies must develop a coherent and robust systems of regulation and control that can respond effectively to new challenges.

Keywords: muscle foods; modified atmosphere packaging; CO; shelf-life; best practice; regulation

1. Introduction

Oxidative browning is the primary basis for consumer rejection of fresh beef in retail display. The meat industry has made a great effort to develop techniques that can improve color stability. Consumers prefer that raw meat of good quality is bright red, which is an indicator of freshness [1].

Traditionally in developed countries, fresh meat is wrapped in O_2-permeable film allowing the meat to turn bright red. This bright red color is retained under these conditions for about a few days (~3 days). The shelf-life of perishable fresh meat is limited in the presence of normal air. Chilled storage will significantly reduce the rate at which detrimental changes occur in the food, but will not extend the shelf-life sufficiently for retail distribution and display purposes.

Over the past decade, the use of case-ready modified atmosphere packaging (MAP) has increased by the meat industry in various countries. Carbon dioxide (CO_2), nitrogen (N_2) and oxygen (O_2)

are the gases most commonly used in MAP fresh meats. Indeed, the majority of red meat products are packaged in a high O_2 environment (~80% O_2) to reduce myoglobin (Mb) oxidation and provide a stable, attractive, "bloomed" red meat color, in a proportion of at least 20% CO_2 to prevent the growth of Gram-negative bacteria responsible for aerobic spoilage such as *Pseudomonas* spp. However, high O_2-MAP (HiO_2-MAP) can increase lipid and protein oxidation, with negative effects on meat flavor [2,3] and texture, which reduces the tenderness and juiciness of the meat [4,5]. Another concern of HiO_2 packaging is the possible development of premature browning (PB); the phenomenon develops when meat is cooked, resulting in meat that appears done before it has reached a temperature that renders it microbiologically safe [6–9], thus causing the risk of consumption of undercooked meat with pathogenic bacteria.

To extend red color stability and avoid the drawbacks of aerobic packaging, an anaerobic MAP technology with 0.4% CO (CO-MAP) was approved [10] for use with fresh meats in the USA [11]. Previous studies have proven that CO can significantly increase the color stability of beef compared with other packaging methods. The possible reason that CO-MAP could enhance red color stability was related to the higher stability of carboxymyoglobin (COMb) than oxymyoglobin (O_2Mb) [12,13], owing to the stronger binding of CO to the iron-porphyrin site on the Mb molecule [14]. The main advantages of CO include the maintenance of the desirable attributes mentioned above in relation to color stability, growth reduction of spoilage organisms and prevention of oxidative processes [11]. However, due to the potential toxic effect of CO, its use is controversial in some countries.

Vacuum packaging (VP) is another common method used to distribute meats and in supermarkets for their retail/display. Storage of beef in gas-impermeable packages confers to product a purple color owing to the formation of deoxymyoglobin (DeoxMb). The anaerobic environment delays aerobic microbial growth such as *Pseudomonas* spp. and psychrotrophic aerobic bacteria and improves microbial shelf-life. However, consumers prefer the appearance of bright red beef compared to the darker vacuum-packaged beef products [15]. It is possible that a pre-treatment with CO will make it possible to overcome the unattractive color in the vacuum-packed meat pieces. This will maintain an attractive red color throughout the exposure and sales period, allowing for a more tender meat due to optimum ripening during this period.

Since 1985, Norwegian meat industries have used 0.4% CO in MAP of muscle foods (fresh beef, pork and lamb) with 60–70% CO_2 and the balance as N_2. About 60% of the fresh meat in Norway has been sold using this gas composition [16]. The use of CO for MAP of meat has been prohibited in Norway since 1 July 2004, due to the implementation of European Union (EU) food regulations.

The scope and purpose of this paper and the controversy about the use of CO in meat packaging were discussed. The use of CO in fresh meat packaging gives promising results due to its positive effects on overall meat quality, enhancement red color, reduced lipid oxidation and microorganism growth inhibitions, which result in shelf-life prolongation during wider distribution of case-ready products. However, in realistic concentrations, CO as such has no antimicrobial effect, and CO_2 in sufficient concentrations is required for delaying the growth of Gram-negative bacteria.

The use of CO in the food industry is controversial. Some countries approve the application such as the U.S., Canada, Australia and New Zealand, while the EU member states ban it from food processing. CO has previously been reported to mask meat spoilage, and this was the primary concern raised for the prohibition as this may mislead consumers. Another consideration is that the application of CO in meat packaging was not considered feasible because of possible environmental hazards for workers.

Risk of CO toxicity from the packaging process or from consumption of CO-treated meats is negligible. Moreover, the addition of CO pre-treatments prior to VP may be beneficial to allow a desirable color to be induced while allowing aging to occur within the package and increasing meat tenderness. Additionally, CO is not present in the pack during storage. Several authors do not consider CO only as a toxic gas for our organism. It was quickly discovered that CO is produced endogenously as a cellular protectant by nearly every cell in our bodies when they are subjected to situations of oxidative stress or injury. Although there is increasing interest in the use of CO-MAP for fresh beef, the debates during the last few years concerning the use of CO in meat packaging have not seriously taken into account the preferences of consumers [17]. However, several studies on the attitude of European consumers regarding the use of CO for meat packaging have reported positive relationships that suggest its future potential within the EU. Facilitation of information can help to develop future policies to ensure consumer protection, and therefore, the debate over the use of CO as a protective gas in meat packaging within the EU could be re-considered. This review provides not only the CO-MAP technology as the solution for the shelf-life issues of muscle foods, but also the new controversy in the use of CO in meat and fish packaging, with the key arguments of both parties. The impact of this review to the field of academic research, food industries and public health was also discussed.

2. Fresh Meat Packaging Methods

2.1. *Raw Meat Spoilage-Associated Storage Conditions*

Fresh meat suffers during refrigerated storage some modifications, which can be either physical (water loss) or chemical (color and odor modification) or microbiological. The shelf-life of fresh meat is not unlimited. As is known, its alteration is due to a greater or lesser extent to the presence of atmospheric O_2, as a consequence of a series of well-known mechanisms: 1. the oxidizing chemical effect of the atmospheric O_2; 2. the growth of aerobic spoilage microorganisms; 3. photo-oxidation. All of these factors, either alone or in combination, can result in detrimental changes in the color, odor, texture and flavor of meat. Maintaining meat's quality attributes throughout its shelf-life has been a perennial challenge for the meat industry. In this context, the best packaging methods in combination with low temperatures have been considered an important technology with respect to maintaining quality standards with optimal distribution and extending shelf-life for the retailers.

2.1.1. Microbial Spoilage

Fresh meat is an excellent source of nutrients and makes it an ideal environment for the growth of spoilage microorganisms and common pathogens. Therefore, it is essential that adequate preservation techniques are applied to maintain its quality and safety. The presence of aerobic conditions results in the growth of mainly aerobic psychrotrophic bacteria types, *Pseudomonas*, *Acinetobacter* and *Moraxella*, responsible for fresh meat spoilage during aerobic cold storage. It has been reported that microbial spoilage of meat occurs when counts of aerobic bacteria reach levels of 7 $\log_{10}$ cfu/g [18]. This level is commonly found to be correlated with sensory deterioration, like off-odors and the presence of slime on the surface of meat.

2.1.2. Lipid Oxidation

Lipids are an important component of meat and contribute to its several desirable characteristics. Meat oxidation not only influences the eating quality of the products, but also has harmful effects on the health of humans by the formation of carcinogenic substances. Malondialdehyde (MDA), which is a degradation product of lipid oxidation, has been criticized as a carcinogenic factor in food. The development of rancidity in meat by lipid oxidation begins at the time of slaughter and is strongly enhanced during processing and storage during which the phospholipids are released from the membrane so they can be more easily oxidized. These processing steps disrupt muscle structures, causing unsaturated fatty acids to react with atmospheric O_2 and to increase contact with endogens macromolecules that promote auto-oxidation in the meat systems. Lipid oxidation might not be considered a limiting factor for beef shelf-life, as it occurs at a slower rate than pigment oxidation or microbial spoilage.

Free radical chain reaction is the mechanism of lipid peroxidation and reactive O_2 species (ROS) such as hydroxyl radical and hydroperoxyl radical and iron are the major initiators of the chain reaction in the development of lipid peroxidation in meat and meat products.

The formation of volatile lipid oxidation products strongly reduces the consumer's acceptability of the product in a number of ways, including off-odors and off-flavor formations. Oxidative processes can also affect the ability of the membranes to hold water and may contribute to drip loss and consequently cause changes in the functional and sensory characteristics of the meat [19]. Therefore, the development of lipid oxidation varies due to the livestock species and environmental conditions.

There are, thus, many factors that influence lipid oxidation and off-flavor in muscle foods. The oxidation rate will depends strongly on the presence of O_2, the pro-oxidants/antioxidants balance, the degree of unsaturated fatty acids and the storage conditions. Control of these factors is the best way to retard lipid oxidation and off-flavor in meat products. Antioxidants and chelating agents are the most effective inhibitors of lipid oxidation [20,21].

2.1.3. Pigment Oxidation

Due to the rejection of fresh discolored beef in retail/display, the meat industry could suffer financial losses estimated at several millions of dollars and according to Mancini and Hunt [13], and a 15% price reduction of retail meat is reported due to surface discoloration. The redness of meat color is produced by oxygenation of DeoxMb to O_2Mb, due to exposure to O_2. This oxygenation is considered reversible according to the partial pressure of O_2 (pO_2). Discoloration of the meat surface results from oxidation of pigment to metmyoglobin (MetMb). Often, the oxidation of pigment is only slowly reversible by enzyme-mediated reduction of MetMb in the early stages of pigment oxidation (Figure 1). Therefore, preservation of meat color appearance involves primarily the prevention or slowing of MetMb formation on exposed meat surfaces.

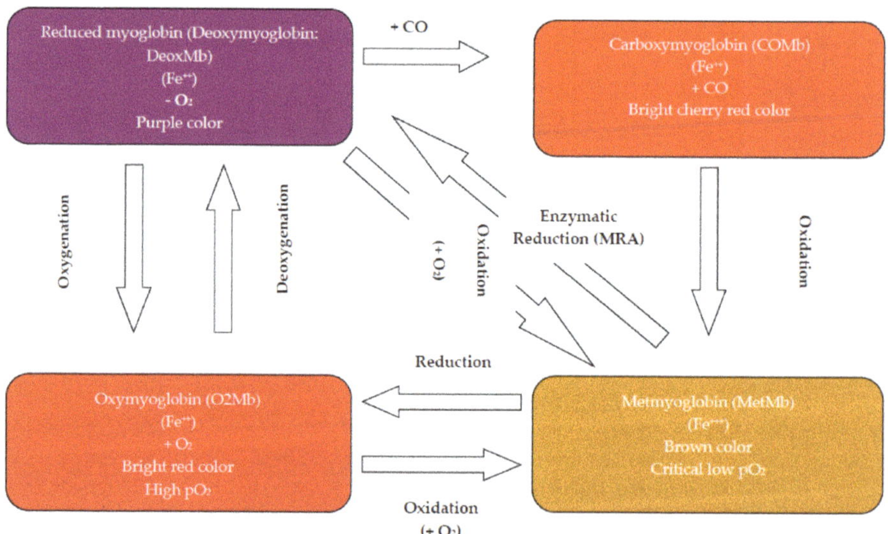

Figure 1. The cycle of color in fresh red meat.

The heme group of Mb contains an iron atom within a protoporphyrin ring. The color that Mb imparts to meat is determined by the redox status of its heme iron and the chemical species bound to heme. DeoxMb is most commonly observed in fresh beef slices just after cutting from whole muscle. The relative concentration of O_2 to which Mb is exposed is critical to the redox form it will assume. When O_2 is present at approximately 4–6 mmHg, the MetMb form will predominate [22]. This is a very important relationship and must be considered in light of natural O_2-consumption that may occur from mitochondrial activity, lipid oxidation and/or bacterial growth. Overall, studies demonstrated that O_2 consumption concomitant with *P. fluorescens* growth decreased pO_2, which accelerated O_2Mb oxidation and consequently beef discoloration. Meat appearance in retail/display is influenced by several factors such as species and muscle location, marbling, aging period, display lighting and temperature [23]. Hydrogen sulfide (H_2S) is presumably produced by contaminating bacteria, binding heme and forming the green pigment, sulfmyoglobin (SHb). Lipid oxidation is positively correlated with pigment oxidation [18,24–26]. The correlation of lipid oxidation with a decrease in redness is related to the formation of precursors for O_2Mb oxidation from lipid oxidation, and ferric hemes are believed to promote lipid oxidation. Generally, consumers discriminate beef containing 20% MetMb on the surface [27], and it is rejected from purchase when MetMb levels exceed 40% [28]. Microbial contamination could play a role in the phenomenon of meat surface discoloration by competition for O_2 and possibly reduce pO_2 below the critical level for oxidation of pigment to MetMb [29].

2.1.4. Photooxidation

In retail cases, displaying meat under lights accelerates the formation of MetMb, which produces unattractive brown color. The effect of light on lipid and pigment oxidation has been demonstrated in meat systems [28]. UV-light (UV-A) is more effective than visible light in inducing oxidation of lipid and pigments [30,31].

Discoloration of fresh beef is significantly faster upon light exposure than when kept in the dark, an observation of particular importance for retail display [28]. The dull appearance associated with light-induced Mb decay has been recognized for decades [32], but only a few mechanistic details have been elucidated. The fluorescent tubes usually used in supermarkets for meats' retail/display have emission spectra designed to maintain a color balance. In a study on the effect of different sources of light on packaged meat quality, Djenane et al. [28] reported that the level of MetMb on the surface of the meat displayed under the standard tube (Mazdafluor Aviva TF/36 w, Philips, Eindhoven, The Netherlands) can reach 70% of MetMb on the 17th day of display, while that exposed to the Promolux®tube (Platinum L36 w, Market Group Ventures Inc., Shawnigan Lake, BC, Canada) or in the presence of a UV filter (polycarbonate) can reach only 40% of MetMb for 28 days of display. Standard fluorescent tubes used in display cabinets may transmit radiation below 400 (UV-A ~390 nm), so it must be taken into account for its deleterious effects on meat display life [33]. Nevertheless, light discoloration depends on many technological factors (temperature, pO_2, meat pH, storage time, free transition metal ions, light wavelengths, type of display lighting, atmosphere) that can affect the photocatalyzed autoxidation [34]. The effect of light on lipid oxidation has been demonstrated in various food systems [35–38]. Lee et al. [39] found that lipid oxidation was influenced by light intensity (1000 vs. 3000 lux). Herbs and spices that are usually added to sausages and processed meats may contain chlorophyll. The latter can absorb light and promote lipids' photo-oxidation.

2.2. Fresh Meat Shelf-Life

Maintaining quality attributes throughout its shelf-life has been a perennial challenge for the meat industry and government agencies (Figure 2). Shelf-life is a frequently-used term that can be understood and interpreted differently. In 1974, the U.S. Institute of Food Technologists defined shelf-life as "the period between the manufacture and the retail purchase of a food product, during which the product is in a state of satisfactory quality in terms of nutritional value, taste, texture and appearance". According to The Institute of Food Science and Technology in the United Kingdom, shelf-life is "the period during which the food product will remain safe; be certain to retain desired sensory, chemical, physical, microbiological and functional characteristics; and comply with any label declaration of nutritional data when stored under the recommended conditions".

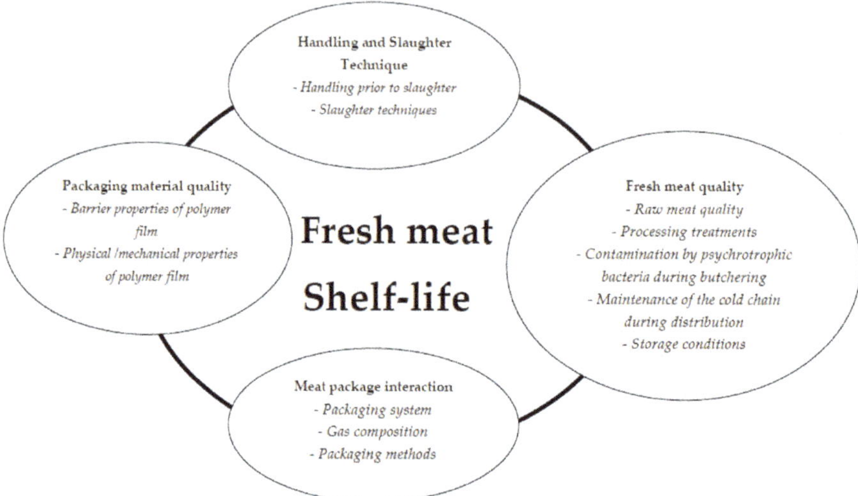

Figure 2. Factors influencing shelf-life of fresh meat.

In the EU, shelf-life is not defined in law, nor is there legislation about how shelf-life should be determined. According to EU regulations (Directive 2001/95/EC of the European Parliament and of the Council on General Product Safety), the manufacturer is responsible for putting safe products on the market. More recently, fresh beef shelf-life was determined based on sensory analysis, chemical, physical and microbiological properties. Shelf-life was therefore defined as "the period between slaughter of animal and the simulated retail purchasing, during which the meat retains all its qualities attributes" [18]. However, in commercial practice, this definition overlooks the fact that the consumer may store the product at home for some time before consumption.

2.3. Packaging Options

The role of centralized packaging systems in the longer modern food supply chains is being increasingly recognized as it has multiple functions and is very important in terms of increasing product shelf-life by retarding food-quality degradation and ensuring food safety [40]. The four major packaging systems include vacuum packaging (VP), vacuum skin packaging (VSP), modified atmosphere packaging (MAP) and polyvinyl chloride over-wrap film (PVC). The four approaches differ in their preservative capabilities and their applicability to the centralized packaging of retail meat. Moreover, fresh meat commercialization strategies have notably changed during the last few decades. Therefore, maintaining quality and color appearance is fundamental during the distribution and marketing of meat. For this reason, packaging innovation has become indispensable to increase the shelf-life of this product.

2.3.1. Emergence of MAP Case-Ready Meat Products

During the last decade, case-ready meat packaging techniques continued to grow. Traditionally in the industrialized countries, fresh meat is wrapped in oxygen-permeable polyvinyl chloride film. The advantage of this technique allows rapid oxygenation of the surface pigments and development of the red bright color, but oxidation of pigment occurs within a few days (1–7 days). Over the past decade, the use of case-ready meats has increased in the U.S. and developed countries' markets. However, as a result of an increasing demand for fresh and ready-to-use products, a need has emerged for further studies involving the possibility of extending the shelf-life of the refrigerated meat. In fresh meat, Mb can exist in several redox states. Understanding pigment chemistry has promoted central packaging of case-ready meat products, which can increase color shelf-life via the use of MAP and other technologies. These products are cut into consumer-ready portions, packaged and typically sold under labeled names that consumers can accept or reject when making purchasing decisions.

Commercial retailing of fresh meat packaged under modified atmosphere (MA) tray systems was introduced in the early 1970s. Case-ready modified atmosphere packaging (MAP) can reduce the costs of fabrication and packaging at retail outlets and other benefits; like preventing from being out-of-stock. The most common gas mixture used for MAP fresh beef is 80% O_2 and 20% CO_2. High O_2 packaging could increase color stability up to 21 days [19] compared to 4–7 days for meat packaged in O_2-permeable PVC film over-wrap [20]. However, at purchase point, meat initial microbiological load, meat temperature history, intensity of light, display period, temperature, location in the display case and characteristics of packaging materials frequently determine the effectiveness of preservative packaging.

2.3.2. High O_2 MAP

MAP is based on modifying the composition of gas in contact with food by replacing the air with a single gas or a mixture of gases (gas naturally present in the atmosphere). The major gases in dry air by volume at sea level are N_2 (78%), O_2 (20.99%), argon (0.94%) and CO_2 (0.03%), but the percentages vary when calculated by weight [20]. The use of MA is not a new concept in food preservation. In the 19th Century, scientists discovered that high levels of CO_2 showed antimicrobial effects. Since the 1930s, the Australian meat industry was using CO_2 atmospheres to extend shelf-life of fresh meat exports, but for a lack of auxiliary means, MAP processing was replaced with freezing after World War II due to lower costs and longer shelf-life. During the last few decades, HiO_2-MAP has been and continues to be widely used in case-ready meat production. HiO_2-MAP promotes the desirable bright red color of meat during storage and display due to Mb oxygenation. The most common gas mixture is 60–80% O_2 and 20–30% CO_2 [41], even though it is demonstrated that a minimum of 55% O_2 is sufficient to maintain a good meat color [2]. Even though a limiting value of about 5% O_2 partial pressure is needed to maintain O_2Mb [22], O_2 higher than 13% will provide predominant O_2Mb pigments [42]. This is readily achievable with air-permeable overwrap packaging or HiO_2-MAP.

Gases Used in MAP

The conventional gases used in MAP are N_2, O_2 and CO_2. The choice of gas used singly or in combination depends on the packaged product. Today, packaging films are available with different gas permeabilities to meet the different requirements of the food industry. CO_2 is present in the atmosphere at a low level (0.03%). CO_2 is a colorless gas with a slight pungent odor at high levels. The solubility of CO_2 in water and lipid phases of the product increases with decreasing temperature [43]. For this reason, the antimicrobial activity of CO_2 is markedly greater at temperatures below 7 °C. This has significant implications for MAP of foods.

The antibacterial effects of CO_2 have been known for decades. CO_2 is effective against psychrotroph bacteria and has high potential for shelf-life extending of chill-stored food [44]. Gram-negative bacteria were considered more sensitive to CO_2 than Gram-positive bacteria. However, various authors indicate that high levels of CO_2 (>20%) result in undesirable brown color in meats [45]. The antimicrobial mechanism of CO_2 could be explained by its ability to penetrate the bacterial membrane, causing a great intracellular pH change [46]. In general, CO_2 increases the lag phase and generation time of spoilage microorganisms. Antibacterial effects of CO_2 are markedly temperature dependent, and it is therefore imperative that chill temperature be maintained across the supply chain for health concerns, thus enhancing the shelf-life of the perishable meats. O_2 is a colorless, odorless gas that is highly reactive and supports combustion. It has a low solubility in water. O_2 promotes several types of deteriorative reactions in foods including fat oxidation, browning reactions and pigment oxidation. Most of the common spoilage bacteria and fungi require O_2 for growth. HiO_2-MAP has been and continues to be widely used in case-ready meat production. HiO_2-MAP promotes the desirable bright red color of meat during storage and display and can slow pigment oxidation. However, HiO_2-MAP can increase lipid and protein oxidation and supports the growth of spoilage bacteria with negative effects on meat flavor and texture [47]. Another problem created by exposing beef to high O_2 atmospheres is called "premature browning". N_2 is a relatively un-reactive gas with no odor, taste or color. The low solubility of N_2 in foods can be used to prevent pack collapse by including sufficient N_2 in the gas mix to balance the volume decrease due to CO_2 and O_2 absorption into the product. It is used to displace air and, in particular, O_2 from MAP. Since air and consequently O_2 have been removed, the growth of aerobic spoilage organisms and oxidation reactions are inhibited or stopped. N_2 is an inert gas that is not reactive with meat pigments or absorbed by the meat; therefore, it prevents from collapsing and maintains the integrity of the package by its presence in the headspace. The noble gases such helium (He), argon (Ar), xenon (Xe) and neon (Ne) are characterized by their lack of reactivity. These gases are being used in a number of food applications.

2.3.3. Vacuum Packaging

VP is another common method used to distribute meat and continues to be in many cases the most cost-effective packaging strategy. VP extends the shelf-life of beef even longer than HiO_2-MAP. Although residual O_2 levels of 0.15–2.0% predispose fresh beef products to browning [13] because fresh meat is very susceptible to MetMb formation by low O_2 pressure, it may be necessary to reduce residual O_2 to 0.05% or lower to inhibit MetMb formation and induce optimal re-blooming upon exposure of meat to air. Various O_2 absorbers or scavengers have been used to reduce the concentration of residual O_2. The main desirable effects of VP are inhibition of off-odors [48] and spoilage by *Pseudomonas*, but generally, consumers prefer meat with a bright red color compared to the darker VP beef [15]. Another disadvantage of VP is the purging of the vacuum pack in the folds after air removal, which often leads to increased microbial growth and less attractiveness to consumers [49]. The determination of the barrier properties of a polymer is crucial to estimate and predict the product-package shelf-life during storage and subsequent display of vacuum-packaged fresh beef. The specific barrier requirement of the package system is related to the meat product characteristics.

Vacuum skin packaging (VSP) is an alternative to conventional VP for retail portions. VSP is an advanced type of VP, which helps to avoid the formation of film wrinkles by making the upper film shrink tightly around the meat, and consequently, meat in VSP is considered more attractive. VSP is a technology that has been designed to prevent purge loss while maintaining many of the benefits of VP. In recent years, this packaging technology has grown significantly in industrialized countries due to the demand for a more tender meat with a long shelf-life during retail/display. The O_2 deficiency in the VSP case has the same effects as those of VP; the meat in this case has a dark purple color due to DeoxMb [50]. In recent years, meat marketing efforts have been carried out by large meat industries and large supermarkets to attract a wide range of consumers with respect to the individual cuts of vacuum-packed meat; surprisingly, this approach has not been successful [51].

2.3.4. Safety of MAP

Maintaining the quality and safety of muscle foods from slaughter to consumption is highly important in the modern food supply. Concerns have been expressed about the risk of pathogens in meat packaging under MA. The increase in shelf-life of MAP meats through inhibition of spoilage bacteria may provide sufficient time for stimulated pathogens to grow to dangerous levels while the food still remains attractive to the consumer. Whatever the packaging system, keeping the continuous chill chain throughout all of the storage period is the most important key factor influencing the storage life of fresh meat. MAP of fresh meats is generally considered less hazardous if cooking is correctly carried out. It is also imperative that catering factories and consumers at home maintain adequate refrigeration and, if it is possible, monitor internal temperature during cooking to assure that products reach a high internal temperature to assure destruction of pathogens.

The hurdle concept is widely accepted as a food preservation strategy; its potential, using MAP, has still to be fully realized. When MAP is combined with other preservation methods, its effectiveness may be highly enhanced [52,53].

3. CO in Fresh Meat Packaging

3.1. What Is Carbon Monoxide?

CO is a toxic gas. It is produced by the incomplete combustion of carbon-based materials (fossil fuels, industrial and biological processes, wood, etc.) [11,54]. CO is an odorless, colorless, tasteless gas, non-irritating and non-suffocating. Its density is very close to that of air (0.967). It diffuses very quickly in the ambient environment occupying all the space available, which is potentially dangerous in a closed environment. In the biological medium, it is easily bound by coordination to the divalent iron (Fe^{2+}) or to the copper (Cu^{2+}) of the hemoproteins. In addition to its production by the incomplete combustion of hydrocarbon materials, significant quantities are also produced during the operation of internal combustion engines due to incomplete combustion of the fuel. A small quantity of CO is naturally produced endogenously in humans [55], regulating blood flow and blood fluidity [56]. CO can be produced during irradiation of meat [57–61]. The production of CO in irradiated samples was irradiation-dose dependent [58,62]. The major sources of CO in irradiated meat are amino acids and phospholipids [63]. CO is also produced by reactions between meat components and free radicals produced by radiolysis [58,62,64].

3.2. Health Implications of CO

CO is called the "silent killer" because if the early signs are ignored, a person may lose consciousness and be unable to escape the danger. The first descriptions of CO and its toxic nature appeared in the literature over 100 years ago; recognized since 1895 by Haldane [65]. The ancient Greeks and Romans used CO to execute criminals. Application of CO for sedation and killing of animals has mainly been carried out in scientific studies, but has been practiced at an industrial level for killing of mink. Use of CO for sedation and killing may be important, especially in order to enhance both the animal welfare and the overall meat quality. Utilization of CO for euthanizing of animals may also be carried out by veterinarians at laboratory facilities [66]. In most countries, stunning prior to sacrificing of animals is required. The method to stunning should ensure that the animals reach an unconscious state fast, without any pain, and the animals should not recover consciousness before death. Meanwhile, Kosher and Halal practices of the slaughter of animals without stunning are in use.

The past twenty-seven years have been marked by an explosion in the number and quality of studies regarding the actions of CO in mammalian systems. The toxic action of CO is due to the blockage of the O_2-carrying function of hemoglobin (Hb) through the formation of COHb instead of oxyhemoglobin (O_2Hb) and prevents the body from using O_2. Thus, cells, tissues and vital organs may become hypoxic and undergo irreversible anatomical, biochemical and physiological changes, leading to death and morbidity. Thus, symptoms appear when COHb is >10% [67]. The fetus and infant are the most predisposed to harmful effects of CO compared to adults due to higher metabolism and the presence of fetal hemoglobin, which has a greater affinity for CO than adult hemoglobin [68]. On the other hand, the risk of developing autism in children is also linked to CO exposure [69]. The affinity of hemoglobin for CO is 240-times greater than that for O_2. A small amount of CO (~0.5%) is formed naturally in the human body from the breakdown of hemoproteins [56]. The average COHb level in nonsmokers is 1.2–1.5% (from both endogenous and environmental CO) and 3–4% in smokers [54]. Clinical results showed that children born to women smokers exhibited low intellectual development and poorer performance on cognitive tasks [70]. The half-life of COHb is 4–6 h in the mother, but much longer in the fetus (18–24 h), which accentuates the effects of hypoxia on cerebral functioning and explains the rates still being higher COHb in the fetus than in the mother. Today, we now know that CO has a number of different action sites. More knowledge about the physiological process involving

CO is necessary especially after the discovery of the heme protein neuroglobin (Ngb) in the brain of vertebrates [71]. Neuroglobin is thought to act as a reservoir for O_2 and in that way prolongs the activity of the nervous system. CO reacts with the heme protein Ngb in the brain and might also take part in biological signaling. It is possible that CO has an important function in humans. CO may have a physiological influence on the mind's functioning since CO acts as a biological signal in regulating the cyclic guanosine monophosphate (GMP) and most likely works as a neuronal messenger [72,73]. It was discovered that CO is produced endogenously as a cellular protectant by nearly every cell in our bodies when they are subjected to situations of oxidative stress or injury [74–76]. Recently, CO has emerged as a potential therapeutic agent for the treatment of various cardiovascular disorders [77–79]. CO is a necessary molecule for normal cell signaling and can play a therapeutic role in humans [80]. Kim et al. [81], Onyiah et al. [82], Soni et al. [79] and Steiger et al. [83] established that CO can be used as a potential pharmacological cytoprotective (anti-inflammatory protection) agent against several diseases.

3.3. CO Application in Meat Packaging

The use of CO for meat packaging is not allowed in most countries due to the potential toxic effect, and its use is controversial in some countries. Nevertheless, according to Sørheim et al. [16,54], the method of adding 0.4% CO to a commercial gas blend with 60% CO_2 and 39.6% N_2 for case-ready packaging systems of beef, pork and lamb was developed by the Norwegian meat industry, starting in 1985. The use of CO for meat was discontinued in July 2004 due to pressure from European trading partners [84]. In Norway, the low CO packaging process grew to 60% of the retail red meat market [85]. The commercial success and safety record of the Norwegian process was a factor in the renewed interest in fresh meat packaging using CO in the United States. CO has a long history of application within the meat industry for its color-stabilizing effect coupled with its antioxidant abilities. Most CO-modified atmospheres contain no O_2, which limits the oxidation and growth of aerobic microorganisms. CO binds to the sixth coordinate of the heme group centrally located within Mb and forms a bright cherry-red color (COMb). The affinity of DeoxMb for CO is 28–51 times greater than for O_2 [86]. MbCO is more stable than MbO_2, making it is less likely to oxidize to the brown pigment, MetMb, during display [87]. Important findings in extending the shelf-life of fresh meats by MAP alone and by other treatments since 2000 are summarized in Table 1.

Table 1. Important findings in evaluating CO-modified atmosphere packaging (MAP) on extending the shelf-life of fresh meats.

Publication	Results/Conclusions
Luño et al. [88]	The presence of CO and 50% CO_2 extends the shelf-life by inhibition of spoilage bacteria growth, delayed metmyoglobin (MetMb) formation; maintains red color and odor of fresh meat and slows down oxidative reactions. CO concentrations of 0.5–0.75% were able to extend shelf life of packaged fresh meat by 5–10 days at 1 °C.
Nissen et al. [89]	At 4 °C, the shelf-life of ground beef packed in MA, based on color stability and background flora development, was prolonged for the high CO_2 (60%)/low CO (0.4%) mixture compared to high O_2 packaging (70% O_2/30% CO_2), but at 10 °C (abuse temperature), the shelf-life was <8 days for both packaging methods. The growth of *Y. enterocolitica* and *L. monocytogenes* in ground beef stored in the high CO_2/low CO mixture was not increased as a result of prolonging the shelf-life. However, the growth of strains of *Salmonella* at 10 °C in this mixture does emphasize the importance of temperature control during storage.
Sørheim et al. [85]	By adopting the use of CO in combination with high CO_2 for meat packaging under MAP, retailers must adopt packaging systems indicating the deadlines for optimal use of the packaged product. However, as with other perishable foods in MAP, food products must be handled according to strict hygienic standards, and low storage temperatures must be maintained in a continuous chill chain.
Jayasingh et al. [90]	Ground beef packaged in 0.5% CO would maintain color stability for several weeks. The penetration of CO and depth of formation of COMb in the meat is dependent on the concentration of CO in the atmosphere, the time of CO exposure and the structure of the meat. For safety issue concerns, the workers were not exposed to dangerous levels of CO during MA packaging, which was verified by CO detectors.

Table 1. *Cont.*

Publication	Results/Conclusions
Kusmider et al. [91]	The low levels of CO (<1%) incorporated into MAP will maintain a stable, cherry-red color along with extended shelf-life of irradiated ground beef during 28 days of storage, thus countering the potentially negative color effects of irradiation.
Krause et al. [92]	0.5% CO significantly improved color stability and sensory attributes for both injected and non-injected pork chops. The depth (bright red band: COMb) of CO penetration from the surface increases as exposure time increases. The depth of the COMb layer steadily increased from the surface to the interior of the chops during exposure to CO. 0.5% CO packages increased in penetration depth from 5 mm on Day 1 to about 10 mm at 14 days, 15 mm at 28 days and 25 mm at Day 36.
John et al. [93]	Raw ground beef packaged in 80% O_2 maintained desirable bright red color until 10 days, but began to darken by Day 14 and lost all red color by Day 21. However, ground beef stored in highO_2-MAP was very susceptible to premature browning (PB) during cooking. PB is a food safety concern, because the cooked product appears done at temperatures where food poisoning organisms may survive. Raw ground beef held in 0.4% CO remained bright red throughout the 21 days of storage. PB and rancidity associated with ground beef packaged in highO_2-MAP were prevented by packaging in 0.4% CO.
John et al. [94]	Premature browning and rancidity associated with beef packaged in highO_2-MAP were prevented by packaging in 0.4% CO, 30.3% CO_2 and 69.3% N_2.
Mancini et al. [13]	Packaging atmospheres containing high levels of O_2 promote beef bone marrow discoloration. Exclusion of O_2 from MA packages and the addition of low concentrations of CO (0.4%) minimized this discoloration by limiting hemoglobin oxidation through packaging atmosphere and will promote a bright red lumbar vertebrae color for as long as 6 weeks after packaging.
Martínez et al. [95]	The retention of color and odor of fresh pork sausages packaged in MA was better achieved using atmospheres containing low CO_2 concentrations (20%). However, increasing concentrations of CO_2 (60%) promoted Mb and lipid oxidation, despite the better antimicrobial effects promoted by the high level of CO_2. The atmosphere containing 0.3% CO together with 30% CO_2 maintained the red color for 20 days, but failed to keep the fresh odor longer than 16 days, in agreement with its small effect on Thiobarbituric Acid Reactive Substances (TBARS) formation and microbial growth.
Wilkinson et al. [84]	Use of CO in MAP provides sufficient shelf-life extension of at least 8 weeks of refrigerated retail-ready pork chops in a master-packaging system. The inclusion of CO in the master-packs has not inhibited the growth of pathogenic organisms. However, given the stable fresh color of CO-treated meat and the lack of inhibition of pathogen growth by CO, there is concern that CO-MAP under certain conditions may pose a food safety risk. As such, safe refrigeration and handling must be emphasized with this type of product.
Wicklund et al. [96]	Chops packaged in CO-MAP were redder and darker than chops packaged in HiO_2-MAP. Based on sensory attributes, the CO-MAP pork was pinker than the HiO_2 pork after cooking to an internal temperature of 70 °C. CO-MAP chops also experienced less purge loss than pork in HiO_2-MAP, which may have contributed to the increased juiciness perceived by the panelists.
Sørheim et al. [97]	CO can be used as an alternative colorant to nitrite in meat products. A gas mixture containing 1% CO was sufficient for achieving a red/pink color of cooked or fermented meat products. Sausages with CO discolored faster during air and light display than nitrite controls. However, discoloration of CO sausages was reduced by anaerobic storage in darkness, showing that absence of O_2 is a necessity for optimum color formation and stability of these sausages.
De Santos et al. [86]	Enhanced pork chops were packaged in 0.36% CO and stored at 4 °C for 0, 12, 19 or 26 days, displayed for 2 days, then cooked to six endpoint temperatures (54, 60, 63, 71, 77 and 82 °C). As storage time increased, Pork chops packaged in CO-MAP retained their internal pink color even after cooking to 82 °C.
Stetzer et al. [98]	Steaks were packaged in 0.4% CO/30% CO_2/69.6% N_2 or 80% O_2/20% CO_2, stored in the dark for 12 and 26 days and placed in a lighted retail display case. Steaks were visually evaluated by trained panelists. Steaks were cooked for consumer color evaluation. CO had no effect on flavor or acceptability and minimal effects on other characteristics, such as color, sheen and purge loss. If the CO environment provides microbiological stability through storage, it can be expected that the raw product appearance will not differ from steaks in traditional HiO_2-MAP.
Aspé et al. [99]	Beef chops (longissimus dorsi) were pre-treated with 5% CO/24 h, vacuum packed and stored at 2 °C. Chops pre-treated with CO were redder during all of the storage period than controls without CO, and microbial shelf-life was 11 weeks. The pre-treatment did not affect pH, water-holding capacity, drip loss or rancidity of the meat stored in vacuum.

Table 1. *Cont.*

Publication	Results/Conclusions
Mantilla et al. [100]	Color stability of tilapia fillets (*Oreochromis niloticus*) was significantly improved by pre-mortem CO treatment (CO-euthanized tilapia). The color of CO-treated fillets was also retained during frozen storage compared to untreated fillets. Hence, pre-mortem CO treatment could be used as a valuable method for improving the color of tilapia during storage.
Linares et al. [101,102]	The effect of the type of stunning (electrically vs. gas), MA and their interactions on meat quality of suckling lamb of the Spanish Manchego breed was determined at 7, 14 and 21 days of storage. Stunning by CO_2 gas prevented the negative effects that electrical systems have on meat quality in lamb apparent during storage. Furthermore, a low CO (30% CO_2/69.3% N_2/0.7% CO) level could give the best meat quality characteristics, even at 3 weeks of storage in the electrically-stunned group. In addition, in the gas-stunned group, it is possible to obtain a product of better color and more tenderness with a post-packing life of 7 days and possibly 15 days using CO in the gas mixture.
Grobbel et al. [103]	Steaks packaged in HiO_2 MAP discolored faster and to a greater extent than steaks packaged by vacuum package (VP) or ultra-low O_2 with CO (ULO_2CO) MAP. Non-enhanced muscles packaged by VP and ULO_2CO MAP had more stable display color and very desirable tenderness and flavor compared with those packaged in HiO_2 (80% O_2/20% CO_2).
Ramamoorthi et al. [104]	The combined irradiation with CO-MAP showed that, after 14 days of storage, aerobically-packaged beef was visually greener and less red than CO-MAP packaged beef. CO-MAP preserved color until 21 days of storage. CO-MAP could be also used to preserve color of beef irradiated at sufficient doses (~2 kGy) to reduce microbial loads to safe levels during 28 days of storage.
Mancini et al. [105]	Packaging steaks in CO (0.4% CO/30% CO_2/69.6% N_2) did not counteract the darkening effects of lactate enhancement. Nevertheless, CO improved color stability of beef steaks compared with high-oxygen packaging (80% O_2/20% CO_2).
Fontes et al. [106]	Fresh blood saturation with CO produces a dried blood of a pleasant pinkish-red color after 12 weeks of storage when packed in low O_2 transmission rates (OTR) bags, with great potential as an additive in meat product formulations.
Raines and Hunt [107]	Increased CO concentration in combination with reduced headspace volume has a greater influence on COMb development. Smaller headspaces with higher concentrations of CO (i.e., 0.8% vs. 0.4% CO) optimize the package size while maintaining or improving the appearance of beef packaged in CO-MAP without compromising consumer safety. This would result in greater efficiency of case-ready meat distribution, making the CO-MAP system more economically feasible and advantageous.
Jeong and Claus [108]	The color of CO-packaged ground beef upon opening the package deteriorated with display time and became less red. However, the initial rate of color deterioration was faster in vacuum-packaged ground beef when it was opened compared to CO-MAP-packaged product. When a CO-packaged product is opened, this color deterioration would provide consumers with a visual indicator of freshness.
Bjørlykke et al. [109]	CO could increase animal welfare when used to slaughter salmon or other fish. Exposure of fish to CO also could improve the quality of products.
Suman et al. [110]	The incorporation of chitosan increased the interior redness of ground beef patties stored in CO-MAP (0.4% CO + 19.6% CO_2 + 80% N_2). This incorporation was also minimizes premature browning (PB) in patties stored under CO-MAP systems instead of under high-O_2 MAP.
Ramamoorthi et al. [111]	Use of CO in MAP gasses has the potential to allow beef subjected to low doses of irradiation to retain its color.
Pivarnik et al. [112]	Filtered smoke (FS) presumably containing high % CO has been used to preserve taste, texture and/or color in tuna (*Thunnus albacares*). Therefore, a general statement indicating that FS treatments would extend shelf-life of tuna in the studied ways of storage: room temperature (21–22 °C), refrigerated (4–5 °C) and iced (0 °C).
Venturini et al. [113]	Packaging under 0.2% CO increased the color stability of beef steaks and ground beef for 28 days at 1 °C, even with residual O_2 concentrations that are considered excessive for anaerobic packaging systems (above 0.1%). After 28 days of storage under CO-MAP and 24 h of air exposure, beefsteaks and ground beef maintained an acceptable appearance and a visual color similar or superior to that of fresh meat. However, after 24 h of air exposure, both the appearance and the smell of steaks and ground beef were considered "slightly unpleasant".
Lavieri and Williams [114]	The CO-MAP (0.4% CO) treatment had no effect on maintaining the COMb "cherry red" fresh meat color during meat spoilage. No potential health hazards or deceptions were revealed due to simultaneous onset of spoilage and the presence of COMb "cherry red" fresh meat pigment in the CO-MAP. The CO absorbed in the meat ranged from 0.22–0.46 ppm CO/g of meat on Day 0 and increased to 2.08–2.40 ppm CO/g of meat on Day 25. The maximum level of CO detected in the meat in this study was below the Environmental Protection Agency (EPA) National Ambient Air Quality Standard of 9 ppm.

Table 1. *Cont.*

Publication	Results/Conclusions
Liu et al. [115]	The CO-MAP (0.4% CO/30% CO_2/69.6% N_2) significantly increased red color stability of all muscles. Steaks in CO-MAP maintained a higher MetMb reducing activity (MRA) compared with those in HiO_2-MAP during storage. After opening packages, the red color of steaks in CO-MAP deteriorated more slowly compared with that of steaks in HiO_2-MAP.
Concollato et al. [116]	CO-treated fish resulted in an earlier onset of rigor mortis, lower final post-mortem muscle pH and higher drip loss after filleting. The assimilation of CO by Atlantic salmon's muscles, through injection in the water, slightly increased lightness (L*) and yellowness (b*) values, limited however to the fresh samples. No significant difference in redness (a*) at any considered time was found between CO and the control group, probably because of the content of astaxanthin that may have minimized the color differences amongst the different groups.
Rogers et al. [117]	CO-MAP (0.4% CO, 30% CO_2, 69.6% N_2) exhibited more desirable color and consumer acceptability throughout lighted retail display of ground beef during 20 days.
Pereira et al. [118]	Addition of CO-treated blood allows the production of better-colored sausages (mortadella) having lower residual nitrite levels.
Fontes et al. [119]	Saturated porcine blood with CO (99%) could substitute meat by up to 20% for mortadella's processing. Therefore, from the nutritional point of view, meat replacement with up to 20% of CO-treated blood is nutritionally adequate for being used in sausage production.
Yang et al. [120]	Aerobically-packaged beef steaks exhibited a bright-red color at the first 4 days. However, discoloration and oxidation became major factors limiting their shelf-life to 8 days. Compared with aerobic packaging, VP extended shelf-life of beef steaks, due to better color stability, together with lower oxidation and microbial populations. Among all packaging methods, CO-MAP (0.4% CO + 30% CO_2 + 69.6% N_2) had the best preservation for steaks, with more red color than other packaging types.
Sakowska et al. [121]	The raw steaks' CO penetration depth increased as exposure times and CO concentration in gas mixtures increased. However, the COMb that formed did not always turn brown during thermal treatment. In cooked samples treated with 0.3% and 0.5% CO-MAP, a red COMb border was visible at the cross-section, whereas other CO packaging treatments had partial or total browning. To create a red color in raw beef and avoid a red boarder in cooked beef, up to 0.5% CO in VP and only 0.1% for MAP can be recommended.
Lyu et al. [122]	The pretreatment of CO combined with O_3 at certain concentrations can be a promising technique to maintain the quality of beef meats under vacuum during storage.
Van Rooyen et al. [123]	The addition of CO pre-treatments prior to VP may be beneficial to allow a desirable color to be induced while allowing aging to occur within the package and increase meat tenderness. The 5-h CO pretreatment exposure time achieved the desirable color, and discoloration reached unacceptable levels by the use-by date. Therefore, applying 5% CO pretreatments may be a potential solution to current packaging issues within the meat sector for safety and enhancing meat quality. In addition, this anoxic packaging technology should prevent any negative quality issues related to high O_2-MAP packaging.
Sakowska et al. [124]	Using CO significantly increased the brightness and the redness of beef steaks in both CO-vacuum packaging and CO-MAP systems during storage for 21 days. They evaluated the effects of 0.5% CO exposure in two MAP (0.5% CO + 30% CO_2 + 69.5% N_2), as compared with conventional VP, on the quality of packaged beef steaks stored for 21 days at 2 °C. The consumers have the greatest desire to purchase the vacuum-packed steaks after exposure in CO.
Van Rooyen et al. [125]	CO as a pretreatment applied prior to VP or VSP may play an important role in overcoming some of the challenges the meat industry faces. This technology provides a prolonged storage, improves the tenderness of the meat and prevents the negative problems associated with other packaging technologies (reduces the risk of the cross-linking/aggregation of myosin due to Hi-O_2 MA, decreases energy usage, storage facilities and distribution costs).

The use of CO gives promising results in the primary package of fresh meat due to its positive effects on shelf-life prolongation during wider distribution of case-ready products. This, in turn, would reduce product and economic losses. By using CO in a modified atmosphere, the need for O_2 to achieve a bright color is eliminated, thus the opportunity to eliminate the detrimental product effects that O_2 imparts to the product. El-Badawi et al. [87] published for the first time an article on the use of CO (air + 2% CO) for the packaging of meat. Previous work reported that MAP with 0.4% CO and VP were the most stable packaging systems for ground beef containing 10–30% fat levels [114]. Modified atmosphere packaging with 0.4% CO is recommended for extended storage of fresh meat in a master-pack arrangement such that export to distant markets can be accommodated [85]. The inclusion of CO in MAP is controversial because the stable cherry-color can last beyond the microbial shelf-life of the meat and thus mask spoilage [23].

3.3.1. Color Stabilizing and Shelf-Life Effects

Fresh meat is a highly perishable product due to its biological composition. Meat color is frequently used as an indicator of freshness and wholesomeness, so color plays a critical role in determining consumers' purchasing decisions. Pigment oxidation, and the subsequent browning, is the primary basis for consumer rejection of fresh retail beef (Figure 3). Considerable effort has been deployed by the meat industry to enhance color stability by innovative techniques. The color of meat depends on the concentration of O_2 and the oxidation state of the Mb. The shelf-life of meat is limited by the initial MbO_2 layers formed during the "bloom", the time required for oxidation of MbO_2 to MetMb, and reaches proportions of total MetMb concentrations such that the meat appears dull and eventually brown [126,127]. The MbO_2 is gradually oxidized to form MetMb, and the kinetics of the process is dictated by several factors such as the muscle type, rate of post-mortem pH decline, packaging film, O_2 consumption, display lighting and temperature and the intrinsic MetMb reducing activity of the muscle [128].

Incorporation of CO in the gas mixture can provide a stable, cherry red color to the meat by formation of MbCO, which is more resistant to oxidation compared to the MbO_2 [54,88]. Jayasingh et al. [90] studied the penetration of CO and development of COMb layers in whole beef muscle and ground beef depending on exposure time. In a gas blend with 0.5% CO, layers of 2-mm COMb developed in beef muscles after 10 h of exposure. Under the same CO exposure, layers in ground beef were deeper, between 7 and 11 mm. Woodruff and Silliker [129] reported that a concentration of 10% CO can penetrate 0.63–0.94 cm beneath the surface of meat, forming a bright stable red COMb layer. Raines and Hunt [107] studied the effects of headspace volume and CO% on COMb layer development in packaged beef steaks and found that the concentration of CO in a smaller headspace resulted in a thicker COMb layer compared with lesser concentration of CO in a larger headspace. Maintaining meat with an attractive bright cherry-red color during retail display is a challenge for processors and the retail industry. Numerous studies have reported the effects of CO on extended color stability, and additional benefits of the application of CO have also been reported. For its enhancement of meat quality attributes and color stabilizing effects, CO has many value-added benefits in meat packaging. El-Badawi et al. [87] conducted one of the first experiments using CO in packaging of fresh beef (2% CO and 98% air) and found that red color was stabilized and maintained for 15 days at 2–3 °C. Seven years later, the same findings were confirmed by Clydesdale and Francis [130]. Jayasingh et al. [90] reported that steaks and ground beef maintained red color for eight weeks when packaged under CO. The possible reason that CO-MAP could enhance red color stability was related to the higher stability of the COMb than O_2Mb [12,13], but other potential mechanisms have not been explored yet. Krause et al. [92] found that CIE a* values (redness) of pork chops were significantly higher in CO (0.4%) atmosphere than in those in traditional aerobic packages and that redness was retained throughout 36 days of storage. Luño et al. [88] reported similar results in which ground beef and beef loin steaks packaged in MAP containing <1% CO retained a stable red color for 29 days. The same authors found that the addition of CO to the MAP gas mixture, either in place of or along with O_2, results in the formation of stable, bright red COMb, and under these conditions, the presence of O_2 is not problematic to meat color probably since CO is able to enhance Mb reduction even in the presence of O_2 (the bright red O_2Mb pigment, arising from exposure of meat to air or high O_2 atmospheres, is known to destabilize during storage) [11,14] and because the affinity of Mb for CO is 30–50-times greater than its affinity for O_2 [131]. Additionally, Liu et al. [115] reported that 0.4% CO-MAP packaging systems can maintain higher MetMb reducing activity (MRA), which is linked to increased color stability, compared to high-O_2 MAP. In contrast, the addition of 0.4% CO to either 20% or 80% O_2 atmospheres did not increase the MetMb reducing activity in five beef muscles [132].

Recently, Sakowska et al. [124] found that CO-MAP packaging systems significantly increased redness of beef steaks during storage for 21 days. Sausages packaged under CO discolored faster during air and light display than sausages treated with nitrite. However, discoloration of CO-packaged sausages was reduced by anaerobic storage in darkness, showing that absence of O_2 is a necessity for optimum color stability of these products [97]. Previous studies have proved that CO can significantly increase color stability of beef compared with other packaging methods. Steaks in 0.4% CO within a master bag achieved 21 days of desirable red color compared with steaks in 80% O_2, which had the highest TBA values [94]. The color stability of beef strip steaks packaged in a 0.4% CO anoxic atmosphere is greater than the color stability of beef strip steaks packaged in HiO_2-MAP [98]. More consistent color stability of beef steaks was obtained by 0.4% CO use (35% CO_2/69.6% N_2) when compared with HiO_2-MAP (80% O_2/20% CO_2) [105]. However, when 1% CO-treated ground beef was exposed to air, the bright red color was lost within a few days [44,133]. Jeong and Claus [108] also found that 0.4% CO-packaged ground beef became less red after opening the package compared with the color of ground beef packaged in VP. However, the factors affecting color loss of beef steaks in CO-MAP upon opening the packages are still not fully understood. Additionally, the variable color stability of CO-packaged meats from different muscles after opening the packages is not clear. The phenomenon of premature browning was evident during cooking of steaks stored in high O_2 [94]. Raw ground beef held in 0.4% CO remained bright red throughout the 21-day storage period. Premature browning in cooked patties was avoided by use of this packaging system [93]. Beef that was treated with 100% CO/3 h before freezing, maintained a bright red color during frozen storage for 90 days [134].

Meat products are commonly manufactured with the addition of nitrite or nitrate. An important function of these ingredients is to create a stable red to pink color of these products. The disadvantage of nitrite is the formation of carcinogenic nitrosamines in nitrite-treated meat products and in vivo by consumption of the products [135]. Sørheim et al. [97] found that CO can be used as an alternative colorant to nitrite in these products. A gas mixture containing 1% CO was sufficient for achieving a red/pink color. Low CO packaging also improves the color life of irradiated ground beef to complement the greatly improved microbial shelf-life achieved by the irradiation process [91]. The stability of meat color is closely related to intrinsic factors of muscle type and metabolism and extrinsic factors of packaging type [136,137]. Each muscle has a unique fiber type and metabolic function. Color stability is also affected by the inherent O_2 consumption rates, oxidation-reduction potential, MetMb reducing capacity and MetMb reductase activity of muscles [136–138].

The shelf-life of perishable meats is limited in the presence of normal air by two principal factors: the chemical effect of atmospheric O_2 and the growth of aerobic spoilage microorganisms. Chilled storage will slow down these undesirable factors, but will not necessarily extend the shelf-life sufficiently for retail distribution and display purposes. Studies were carried out to evaluate the effect of high CO_2/low CO MAP on the shelf-life of fresh meat and meat products under MAP conditions [16,54]. The shelf-life of pork chops was extended to more than 36 days in CO-MAP compared with only 28 days in traditional, HiO_2-MAP, 23 days in VP and 7 days for overwrapped (OW) packages [92]. The shelf-lives at 4 °C during storage of retail-ready meat in high CO_2/low CO mixtures of ground beef, beef loin and pork chops were 11, 14 and 21 days, respectively, when stored in 0.4% CO, 60% CO_2 and 40% N_2 [16]. It has been concluded that storage in low CO, high CO_2 atmospheres is effective for extending the storage shelf-life of cuts. The combination of CO and CO_2 in MAP was beneficial in extending the shelf-life of fresh pork sausage [139]. The use of this gas mixture in fresh meat packaging gives promising results due to its positive effects on color and microorganism growth inhibition, which result in the shelf-life extending during the wider distribution of case-ready products [140].

3.3.2. Antimicrobial Effects

In CO-MAP of meat, the effects of low concentrations of CO on microorganisms seem to be of either no or minor importance. CO has been reported to prevent the growth of microorganisms [141].

Combinations of CO with other gases such as CO_2 to control microbial growth provide an excellent opportunity for meat processors to improve shelf-lives of the retail packed fresh meats [23,142]. CO is selectively bacteriostatic for various microbial populations. CO extends the lag phase and slows the growth rate of *Escherichia coli*, *Achromobacter* and *P. fluorescence* [143] at concentrations of 25–30%, while *P. aeruginosa* is unaffected even at these high concentrations. Gee and Brown [44] conducted research using CO and CO_2 together in an MAP system. Beef patties exposed to 1% CO, 50% CO_2, and 49% air had two log lower levels of bacteria/g than controls after six days of storage. Additionally, Clark et al. [144] showed that increasing CO concentrations with the balance gas being N_2 on beef steaks inhibited the growth of psychotropic bacteria, which also had a positive effect on increased odor shelf-life. This result was due to CO having the ability to increase the lag phase and to reduce the log phase. Luño et al. [88] showed that CO-MAP greatly reduced the psychrotrophic bacteria populations including *B. thermosphacta* in beef, although lactic acid bacteria appeared to be unaffected. Viana et al. [145] evaluated pork loins packaged in 99% CO_2 in combination with 1% CO and reported that *Pseudomonas* sp. growth was limited, and psychrotrophic organisms did not reach 10^7 cfu/g until after Day 20. In CO-MAP, meat maintains bacterial levels less than spoilage levels (~7 log_{10} cfu/g) for ~1 month, but acceptable red appearance is maintained for at least eight weeks [90,146]. This emphasizes the importance of the "use or freeze by" dating system established by the USDA for retail sale of meat in CO-MAP. Woodruff and Silliker [129] reported that a concentration of 10% CO can inhibit microbial growth, further preventing odor and slime by-products. Gee and Brown [14] investigated the effects of different concentrations of CO on pure bacterial cultures of *Pseudomonas*, *Achromobacter* and *E. coli* species. It was concluded that 15–30% CO had an inhibitory effect on the growth of bacteria. These levels, however, far exceed the levels legally allowed for use in the packaging of meat products [10].

At 10 °C, CO-MAP has inhibitory effects on *Yersinia enterocolitica*, *Listeria monocytogenes* and *E. coli* O157:H7, but was not as inhibitory against *Salmonella* strains, indicating that chilled storage is important, regardless of packaging method [89]. In a similar experiment, Cornforth and Hunt [11] indicated that CO in MAP inhibited growth of *E. coli* O157:H7 on inoculated meat even at abuse temperatures of 10 °C. Contrary to all these observations, Bórnez et al. [147,148] found that the use of low CO (69.3% N_2/30% CO_2/0.7% CO) in the packaging of suckling lamb meat did not improve the microbial quality of the packed products. It would be interesting to point out that in realistic concentrations, CO as such has no antimicrobial effect, and CO_2 in sufficient concentrations is required for delaying the growth of Gram-negative bacteria.

3.3.3. Other Effects

Meat tenderness is one of the most commonly-used parameters for the evaluation of meat quality by consumers. The tenderization of meat is reduced under HiO_2-MAP because of protein oxidation [149]. Grobbel et al. [103] demonstrated that beef stored in vacuum or anaerobic 0.4% CO atmospheres was tenderer than in HiO_2-atmospheres. Reduced tenderness is one of the major detrimental effects of the commonly-used HiO_2-MAP. Thus, meat texture is a highly important factor for the meat industry to consider during the production of meat. Cross-linking of myofibrillar proteins is believed to reduce meat tenderness by causing a strengthening of the myofibrillar structure in meat stored under HiO_2-MAP. There have been conflicting findings toward the effects of irradiation on meat characteristics [150–152]. It is well known that irradiation can produce safe foods and extend shelf-life by eliminating food-borne pathogens, as well as spoilage microbes. However, irradiation doses required to kill pathogens can cause undesirable changes in meat color, flavor and odor. Kusmider et al. [91] and Ramamoorthi et al. [104,111] suggest that CO-MAP could be used to preserve beef color irradiated at doses sufficient to reduce microbial loads to safe levels during 28 days of storage, thus countering the potentially negative color effects of irradiation. Lactate is a commonly-used injection-enhancement ingredient that stabilizes the color of beef products by minimizing surface color change through the production of a dark pigment that remains stable during retail storage and display [153]. The ability of CO to produce a bright cherry-red color may counteract lactate's darkening effect. Premature browning is a condition in cooked meat in which the inner parts of the meat turn gray/brown and appear well done at a lower temperature than expected around 60 °C, thus causing the risk of consumption of undercooked meat with pathogenic bacteria. The condition is associated with exposure to O_2 and formation of MbO_2 in the raw meat. Packaging in vacuum or atmospheres containing 0.4% CO without O_2 prevented premature browning, both in ground and whole muscle beef [94], as well as enhanced beef [103]. For bone-in meat, the marrow with its content of Hb is sensitive to pigment oxidation. Storage of bone-in beef in an anaerobic atmosphere with 0.4% CO prevented marrow browning [13]. To minimize darkening problems due to the use of blood in food formulations, various solutions have been proposed. In a previous attempt to solve this problem, Fontes et al. [106,119] showed that blood saturated with CO produced a pigment with a stable and desirable color that could allow a greater amount of blood addition to meat products that would not lead to their browning. As a replacement for nitrite, a 1% CO gas blend was used for the storage of meat raw materials for dry cured sausage or flushed through sausage batters that later were cooked to 80 °C [97]. The sausage had a red color after production, equal or more intense than with nitrite, but the color faded upon air exposure.

3.3.4. CO in Other Foods

Fruits and Vegetables

The USA have applied CO within vegetable processing since the 1970s to prolong the shelf-life of iceberg lettuce during distribution [154–156], and it is recommended as a component of modified atmospheres to prolong shelf-life of tomatoes, cauliflower, cantaloupe, strawberries and citrus [46]. CO has been reported to prevent the growth of fungi in various vegetable foods. Kader [155] showed that CO-MAP greatly reduced the *Botrytis* rot on tomatoes, strawberries and grapes and brown rot on peaches. CO also has been included in the MA of marine transport of foods [155], in part to kill mites and other insects [157]. Incorporation of CO in the gas mixture with a slight vacuum can provide a stable color in minimally-processed fruits and vegetables [158].

Low CO-MAP significantly delayed the internal browning and softening during chilled storage of peach fruit, and the synergistic effects were recorded in combination with 1% of chitosan [159]. Recent research suggested that CO could enhance the anti-senescence ability of plant leaf, improve the superoxide dismutase (SOD), peroxidase (POD) and catalase (CAT) activities of plant tissue and reduce the MDA content of fresh cut Chinese rose flower. In addition, CO fumigation could prevent browning and maintain quality of fresh-cut lotus root slices. The inhibiting browning by CO was related to reducing the activities of polyphenol oxidase (PPO) and POD [160]. A grain storage loss is caused by insect infestation. The application of CO could be used to increase the efficiency of CO_2, especially at high temperatures [161].

Seafood and Poultry

The food industry is continuously attempting to develop new technologies aimed at extending the shelf-life of fish products, without modifying their nutritional and sensory attributes. A bright red color is an important quality determinant in seafood, particularly tuna, as the market value is based on this attribute [162]. Tasteless filtered smoke has been used in salmon smoke houses for decades. Some taste and odor components like phenols, carcinogenic compounds and gases are removed by passing the smoke through washing and several filtration steps. The tasteless smoke naturally contains 15–40% CO [163].

The use of CO either alone or as part of a filtered wood smoke (FS) process has been applied to seafood to maintain the desirable and stable color attributes during storage and distribution [112]. In the USA, CO-MAP, up to 0.4%, are used commercially for packaging of meat, while FS containing 30–40% CO is permitted for pre-treatment of fish [97,164,165]. Fillets of tilapia (*Oreochromis niloticus*) were treated with 100% CO for 30 min, vacuum packaged, frozen at −20 °C and thawed. The CO pretreatment increased the redness and microbiological stability of the fillets, while it did not affect pH, drip or thaw loss [166]. CO is widely used as a supplement for ice or refrigeration storage to delay spoilage and extend the shelf-life of fresh fishery products [167]. Exposure of salmon (*Salmo salar*) or other fish to CO could improve quality and welfare when slaughtered [109,116].

It is very important to assess the toxicological risk associated with the extended shelf-life of muscle foods. Seafood products are of high concern; in particular, for histidine-rich fishes, such as tuna, mackerel, sardine, herring and swordfish, the fraudulent use of CO provides an additional risk since histamine (oxidative decarboxylation of histidine), responsible for toxicological effects, can be formed. The color of turkey meat benefitted from the addition of 0.5% CO [168]; but chicken meat contains low levels of Mb, and the color is less affected by CO.

3.3.5. Co Pre-Treatments

CO pre-treatments before VP can be a good alternative for meat retailers. Additionally, the application of CO pre-treatments prior to VP may overcome this issue, as CO is not present in the pack during storage, and therefore, the risk will be minimized when the trays are opened at the household level. Moreover, the addition of CO pre-treatments prior to VP may be beneficial to allow a desirable color to be induced while allowing aging to occur within the package and increasing meat tenderness. Many studies have reported pre-treatment of meat with CO [92,121,123]. In a study of chilled beef steaks, a pre-treatment with CO stabilized the red color of vacuum sealed meat for 4–8 weeks [169]. In a similar experiment, CO pretreatment, followed by VP and chilled storage of beef steaks, gave an initially more red color, but after six weeks, the redness was similar to that of untreated meat [12]; and the aerobic plate counts were lowered by one log after eight weeks of vacuum storage, compared to no CO pre-treatment. A study carried out by Jayasingh et al. [90] investigated the color stability of beef steaks exposed to pre-treatments of CO prior to VP and chilled storage, aiming at synchronizing microbiological and color shelf-life. The 5% CO pre-treatments prolonged the color shelf-life for five weeks, whereas the 100% CO pre-treatment maintained a color shelf-life of six weeks. They concluded that a pre-treatment for 24 h in a 5% CO-MAP was needed to maintain redness after re-packaging in VP for >21 days and was practical for large-scale application. A 5% CO, 95% N_2 pre-treatment of vacuum-packed beef markedly enhanced the red color and is likely to increase its acceptance by the consumer [99]. The only inconvenience arising from this technique lies in the following fact: the red color of CO-treated meats gradually changed to a brownish discoloration when cuts were stored under aerobic conditions [144]. In contrast, meats continuously held in the presence of low CO/high CO_2 in an MAP environment maintain red color for extended periods. The barrier properties of the films used for the packaging of food are crucial. Therefore, knowledge of gas transfer rates through the material is valuable information.

4. Consumer's Perceptions

At the point of purchase of meat, color, price and visible fat are considered key factors, while tenderness, flavor and juiciness are more closely related to meat eating satisfaction. On the choice of consumers, because it is most difficult to evaluate before purchase, and it is not visible and highly variable. The confidence of consumers in meat based only on color as an indicator of spoilage has probably been exaggerated. Although, at the retail level, consumers will purchase brown meat at a discount, which means that color alone does not indicate the degree of freshness related to meat, a study by Carpenter et al. [170] showed that eating satisfaction of cooked beef was unaffected by the color of the raw meat during the purchase. It is recognized that brown color of raw beef was not a definitive indicator of spoilage.

During the last few years, the use of CO in the food industry has been debated strongly and seriously by several public and private organisms. Cornforth and Hunt [11] concluded that the major disadvantages of CO-MAP of red meat were the negative image of CO held by consumers because of its potential toxicity; on the other hand, storage of products under inappropriate conditions and in the presence of CO could potentially mask visual evidence of microbiological spoilage. It is believed that this stable red color may conceal microbiological spoilage and place consumers at risk. However, Hunt et al. [146] and Brooks et al. [171] showed that addition of 0.4% CO to modified atmospheres for chilled beef did not mask spoilage, even when the color stability was increased. Some recent studies indicated that consumer decisions to eat meat are gradually becoming more influenced by nutrition and health considerations. A consumer preference study that was carried out in Denmark, Norway and Sweden demonstrated that Scandinavian consumers preferred beef steaks in low CO-MAP (0.4% CO) [172]. In contrast, high O_2 packaged steaks were described as more well done, as an effect of premature browning during cooking [173]. Currently, consumers are not informed by the package itself regarding use of CO or elevated O_2 levels in the headspace of MAP meats. Due to the lack of consumer understanding of the science and being misinformed about this technology, consequently, to improve consumer attitudes about CO packaging of fresh meat, communications should be designed to not only inform consumers about the use of CO, but also familiarize consumers with the science of this technology. Grebitus et al. [174] have conducted a study on the preferences of U.S. and German consumers towards the ground beef qualities enhanced by CO-MAP. Results show that consumers in both countries have clear preferences for shelf-life extension expressed by cherry red meat color. U.S. consumer's preferred longer shelf-life as long as the technology is understandable. However, when information is provided about the use of CO in meat packaging, the willingness of U.S. consumers to purchase decreases, and the willingness of some German consumers to purchase increases. Therefore, it is clear that preferences are heterogeneous towards cherry red color once consumers have been previously informed about CO-MAP technology.

An increase in personal knowledge and media exposure influenced acceptance of CO-MAP negatively. Such information can benefit the meat industry, which makes decisions about investing in new CO-MAP packaging methods. Countries differ not only with respect to regulations, but also with regard to consumers' attitudes towards new technologies. Similarly, two recent consumer studies were carried out to evaluate whether Polish consumers would accept CO in meat packaging systems. Consumers had a preference and increased desire to purchase steaks packaged after low CO-MAP pre-treatments for its attractive cherry red color. Consumers did not accept untreated vacuum-packaged beef steaks as they were considered the least attractive and desirable [121]. Therefore, the results from these studies show promise for the future potential application of CO within the EU, despite the current EU prohibition of MAP with CO.

Even more, a high percentage of consumers have shown their reticence toward a fresh meat that even though red in color was beyond its use-by date and had a noticeable off-odor when opened at home.

These findings claim that low CO-MAP is not misleading and that the use-by dates on the packages, coupled with sensory attributes, all contribute to the decision making by consumers about when to reject or accept the product, having shown their reticence towards meat that, although red, has passed its expiration date and has a noticeable odor when the meat container was opened in the home's kitchen. These results may reinforce the hypothesis that CO use is not misleading and that expiry dates on packaging, together with other sensory factors, contribute to consumer decision making when preparing the product for the first time. When this kind of meat is handled and cooked properly, they are also safe.

5. Analysis of CO in Food

A variety of methods is available for the analysis of CO in food materials and in blood, including spectrophotometric methods, infrared analysis and gas chromatography. Analysis of CO can be used to control whether muscle food products have been treated with CO or not, despite that they have not been labeled as such. These considerations underline the suitability of this method to detect even small amounts of the CO-Mb adduct in fish and meat tissue, in regard to the fraudulent treatment of muscle foods in the MAP system. The official method for quantification of CO has a problem, in that a part of the CO is lost during the preparation of the food matrix sample [175]. The official laboratories of food control need not only confirmatory methods, but also rapid low cost screening methods for the everyday activity of food control. Due to the potentially harmful effects of CO, there is a need for precise and reliable measurements. The spectrophotometric method (UV-Vis) has been evaluated in terms of its performance criteria by using tuna fish samples. The results have been compared with those obtained using a head space gas chromatographic technique (HS-GC-MS). The CO levels measured in tuna fish samples by UV-Vis are substantially lower than those revealed by HS-GC-MS [176]. A robust and dependable headspace HS-GC-MS method has been developed and evaluated for the determination of CO-treated tuna fish on the basis of its performance [177].

During the last decade, a simple, confirmative method for quantitative determination of CO in commercially-treated tuna and mahi-mahi (*Coryphaena hippurus*) tissues has been reported using gas chromatography/mass spectrometry (GC-MS), following chemical liberation of CO [162]. Gas chromatography equipped with a flame ionization detector (GC/FID) is an efficient technique with low detection limits and high accuracy [100]. The non-dispersive infrared (NDIR) technique measures continuously the level of CO. the NDIR technique is based on the specific absorption of infrared radiation by the CO molecule. CO has an infrared absorption near 4600 nm. NDIR incorporates a gas filter in order to reduce interferences from other gases, operates near atmospheric pressure and detects CO concentration at 0.05 mg/m^3 [178].

Measurement of consumer's CO exposure level from consumption of CO-packaged meats was based on U.S. Environmental Protection Agency National Ambient Air Quality Standards (EPANAAQS) [11], which presumed that the human metabolism of CO would be equal for digestive processes after consumption of CO packaged meat as the metabolic reactions after CO inhalation. The EPANAAQS for CO is exposure to 9 ppm of CO for 8 h [179], for a typical person inhaling 5 m^3 air/8 h [11]. Lavieri and Williams [114] found that the maximum level of CO detected in the CO-treated meat was below 9 ppm. Droghetti et al. [176] revealed that the UV-Vis spectrophotometric method detects exclusively CO bound to the iron (Fe) atom of the heme protein and not the CO dissolved in solution; probably, this technique underestimates the amount of total CO present in solution.

6. Safety Consideration of CO-Treated Meat

Little information exists in the literature on the consumer's exposure to CO-packaged meat. The toxicological aspects of CO used in MAP of meat were reviewed by Sørheim et al. [16], and they concluded that, with up to about 0.5% of CO, no human toxicity was likely. Sørheim et al. [54] and Cornfort and Hunt [11] found that consumption of CO-treated meat is not associated with any health risks, and meat from CO-MAP results only in negligible amounts of CO and COHb in humans. The Norwegian Food Control Authority (NFCA) has not registered outbreaks or a higher frequency of sporadic cases of food-borne diseases linked to such products since 1985. The increased red color stability of meats exposed to CO was recognized more than 100 years ago [180]. However, the application of CO in meat packaging was not then considered feasible because of possible environmental hazards for workers. For safety reasons, gas detectors are necessary in environments in which CO is applied in any form. Nowadays, exposure to CO in an industrial setting (meat industry) is associated with minimal risks, both due to good practice at the working facilities and equipment design. Human environmental exposure to CO varies greatly. In the same order of ideas, Sørheim et al. [54] indicated that the max level of COHb is recommended to not exceed 1.5%. A COHb level of less than 5% in human blood is not associated with any harm to healthy individuals, and the half-life of COHb in individuals is approximately 4.5 h. The same authors indicated that during various decades of low CO-MAP in the Norwegian meat industry, its use was not associated with any risks to workers. Furthermore, charcoal meat grill workers are one of the many occupational groups that are subjected to CO exposure. It was found that the average COHb levels for these people exceed the 5% COHb level recommended by WHO and the National Institute for Occupational Safety and Health (NIOSH) [181].

According to Wilkinson et al. [84], given the stable fresh color of CO-treated meat and the lack of inhibition of pathogen growth by CO, there is concern that CO-MAP under certain conditions may pose a food safety risk because consumers could conceivably purchase meat that appeared fresh, but was in fact microbiologically spoiled. For this reason, the use of CO in MAP is not allowed in the U.S. and EU. The European Commission [182] also raised a valid concern that if CO meat products are stored under inappropriate conditions, for example increased temperature, which can occur during mishandling or transportation, the presence of CO may mask visual evidence of spoilage. Van Rooyen et al. [123] have combined the advantages of CO pre-treated steaks followed by VP with the cherry-red color while ensuring that discoloration occurred by the end of the shelf-life to address consumers' concerns about spoilage being masked. Following this, the authors determined the optimum exposure time with a 5% CO concentration to be 5 h as this allowed discoloration to occur by the end of a 28-day display period (2 °C). Nevertheless, the increase in color shelf-life of CO-packaged beef is not drastic enough to prevent the negative effects of abusive storage temperatures and extended storage times. The same authors have studied the interrelationships between discoloration, microbial spoilage and off-odors to determine whether a stable COMb formed in CO-packaged meat would extend color life beyond the point of microbial spoilage. No steaks having acceptable color had spoilage levels of microbes ($\geq$log 7). Thus, COMb formed from 0.4% CO did not mask microbial spoilage. Furthermore, Eilert [183] reported that CO does not mask spoilage. When beef was treated with 1% CO for three days (resulting in about 30% saturation of Mb with CO) and then exposed to air, CO was slowly lost; there was a high loss of CO upon cooking (~85%) of CO-treated meat [184]. Concerning fishery products, there are no direct health implications from eating CO-treated tuna. However, since CO treatment makes tuna (*Thunnus albacares*) appear fresh, in this respect, it has to be considered that tuna fish is most commonly associated with incidents of histamine intoxication. Histamine can be formed by oxidative decarboxylation of histidine. Similarly, Chow and Chu [185] indicated that there is a major effect of heating on the release of CO from Mb in CO-treated tuna. Regardless of the temperature of heating, the release of CO was 58–82%. Most people as per the present lifestyle preserve animal food stocks at one time; they are unaware that they are exposed to CO, which can create health hazards [186]. The CO level detected in frozen food samples was found to be slightly above 1 μg/g. Recently, an attempt to calculate the COMb in a package atmosphere and very interesting findings have emerged concerning CO and consumer safety. Based on the fact that the typical meat packages have a headspace of 1.5 L and the ambient air quality standard for CO inhalation is 9 ppm/8 h, fresh packaged meat stored in low CO-MAP with a 1.5-L headspace could contain 0.4% CO. Opening one CO-MAP container in an average space (150 m^3) results in an ambient air CO concentration of 0.042 ppm [11]. Assuming that no COMb has developed, opening one CO-MAP container in 0.8% CO with a 0.4-L headspace results in 0.022 ppm CO in ambient air. After 7 days of display, 9100 of the packages with 0.8% CO in a 0.4-L headspace would have to be opened in one room (150 m^3) to meet the EPA standard of 9 ppm CO. Thus, reducing headspace from 1.5–0.4 L and increasing CO from 0.4–0.8% do not pose a consumer safety risk [107]. For low CO-MAP (0.4% CO) with a 1.5-L headspace, opening of 216 packages for the same area would be required to exceed the EPA standard [187], for a typical person inhaling 5 m^3 air/8 h. On the other hand, an assumed consumption of 250 g fresh meat/day could therefore release 0.18 mg CO (~0.018% COHb). If there were a 100% transfer of CO from the gut to the blood, only a negligible amount of COHb would be added to the 0.5% COHb, resulting from endogenous CO production, and ~1.0% COHb formed from inhalation of the contaminated atmosphere by a non-smoker [54]. Realistically, one would consume even less CO per meal because it is known that only 15% of bound CO remains with the meat after cooking [185]. Exposure through inhalation of headspace gas on opening a package of meat with a MAP containing 0.3–0.5% CO would equally contribute insignificantly to the COHb in the blood when compared to the other sources of inhalation of CO [54].

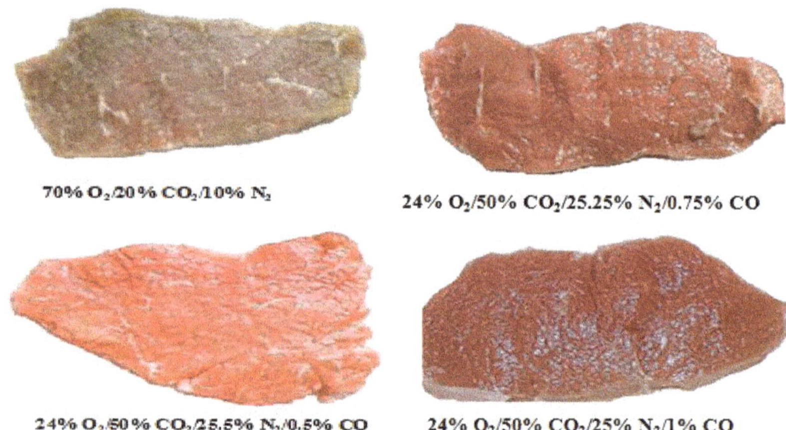

Figure 3. Photos of beefsteaks corresponding to the 21 days of storage under modified atmospheres packaging (MAP) at 1 ± 1 °C [188].

7. Regulatory Status for the Use of CO in MAP Systems

Since 1985, Norwegian meat industries have used 0.4% CO for packaging of muscle foods [16]. The use of CO is controversial. Some countries approve the application such as the U.S., Canada, Australia and New Zealand, while the EU member states ban it from meat processing. The EU considers packaging gases as additives, and CO is not on the list of such gases. Until 2004, low CO-MAP was in use in Norway. Norway is not a member of the EU. Due to trade agreements within the European Economic Area, EU food regulations must be adopted by Norway. In 2002, the Norwegian meat industry applied for an acceptance of low CO gas blends for meat in the EU. The application was positively evaluated by a scientific committee of the EU commission, but eventually, CO was not included on the positive list of additives. The EU imposed a prohibition of CO as a food gas, valid from 1 July 2004. In contrast, an approval of commercial use of low CO packaging for retail meat was announced in the U.S.: the USDA and FDA jointly approved the use of 0.4% CO in an anaerobic MAP master-bag system along with 30% CO_2 and the balance as N_2 during distribution. Further GRAS approval was registered in the U.S. in 2004 [10] for the addition of CO to primary packaging systems, in which the MAP mixture is flushed directly into the meat-containing tray, declaring that CO does not mask spoilage odor during retail meat packaging. New Zealand and Australia also allow low concentrations of CO in centralized packaging systems, and it is considered a processing aid and, therefore, is not required to appear on the product label [82,189]. Similarly, Canada also regulates the application of low CO as a secondary packaging gas under specific conditions in accordance with Health Canada and the Canadian Food Inspection Agency (CFIA) requirements and standards. The European Scientific Committee on Food considered the addition of 0.3–0.5% CO in the MAP system for fresh meat packaging to be safe for human health [183]. However, due to a lack of legal permission, CO is not yet used in the packaging of fresh meat in the European Union. It was recommended to have expiration dates on the label for beef in low CO-MAP; according to Sebranek et al. [190,191], the FDA has examined this issue thoroughly and requires that meat in low CO packaging be labeled with a "use or freeze by" and decided that open date codes for products packed in the CO-MAP system are 35 days following the date of packaging for intact steaks or roasts and 28 days for ground meats. The specific gases do not have to be labeled on the packages. It would be good to point out that meat is subject to natural spoilage processes no matter what type of packaging is used. It is, therefore, important to keep meat at chill temperatures throughout storage, distribution and during retail/display. Consumers should take care to use the meat by the date indicated on the package and practice safe food handling techniques.

8. Could the Application of CO for Meat Packaging Be Re-Considered?

The major factor contributing to retail loss by withdrawal, mainly in industrialized countries, is the discrimination of discolored meat that consumers perceive as unhealthy. In order to reduce these losses and support increased consumer demand and expectation of high meat quality, packaging technology innovations are required. The regulatory aspect concerning the use in developed countries of CO in meat packaging has been a major controversy in recent years. Indeed, the European Union has banned its use in the packaging of meat, mainly because of the fear of masking the deterioration of the packaged product, which could easily mislead consumers. In recent years, great inconsistencies have been recorded between developed countries regarding the adoption of regulations for use of CO in the meat industry, which resulted in obstacles to international trade, limiting export opportunities between countries [17]. The main disadvantage of meat CO-packaging is its possible masking of the microbiological spoilage by formation of the stable, bright red color. For a potential antimicrobial effect, a combination of low levels of CO with high CO_2 concentrations is a better way, and this is important to justify approval by regulatory agencies. Research on pre-treatment with CO followed by VP is promising because meat pre-treatment with CO prior to VP results in improved color while allowing discoloration to occur by the use-by-date, thus addressing safety concerns, assisting in the flexibility in distribution and preventing losses by rejection from the distribution chain. 2002 was the first time that CO was regulated for meat packaging in the U.S. [192]. Two years later, the same federal agency authorized the use of 0.4% CO as recognized as safe (GRAS) in the packaging of meat in case-ready packaging systems [10]. For more than 20 years, Norway has been applying low levels of CO in meat processing systems. However, the Norwegian government suspended its use in 2004 due to the adoption of EU regulations. Australia and New Zealand also regulate low concentrations of CO in centralized packaging systems [189]. Similarly, the Government of Canada also authorized the application of 0.4% CO in red meat packaging. Several studies on the attitude of European consumers regarding the use of CO for meat packaging have reported positive relationships that suggest its future potential within the EU. European consumers preferred beef steaks in low CO-MAP. Despite the current EU prohibition of MAP with CO [124], the decision to ban the use of this gas in the EU should be reconsidered. Sørheim et al. [54] found the application of low concentrations of CO to meat packaging systems to have no toxic effects evident. The European Commission (2001) reported that inhalation of CO-MAP headspace gas containing 0.3–0.5% CO would have no significant effect on the COHb in the blood in comparison to other sources of inhalation of CO [182]. The report also stated that the amount of CO present in fresh meat packaged in low concentrations of CO-MAP is similar to that of the endogenous CO. It is also important to note that during cooking, a considerable amount (~85%) of CO that is bound to COMb and COHb of the packaged meat is lost. In the EU, packaging gases are considered as additives. In order for CO to be approved as an additive within the EU, the following criteria must be met according to Directive 94/34/EEC.

During the last few years, between some European governments within the EU, great efforts have been made to allow the use of CO under controlled and regulated conditions. However, the consensus concerning this fact could take quite a few years for a common agreement. The debates concerning the use of CO in meat packaging have not seriously taken into account the preferences of consumers [17]. CO-packaged meat should be legally labeled for the presence of CO so that consumers can make decisions about their purchases. Furthermore, on the labeling of trays, the consumer must read "the color of the meat does not in any way mean a reliable indicator of freshness". This is to protect consumers and prevent them from being misled about the freshness of the product since CO can maintain an acceptable color beyond the shelf-life period. The pertinent scientific results must be accompanied with transparent and early communications in an effort to ensure that government, consumers and the media have what they need to make thoughtful decisions in the public's interest.

Facilitation of information can help to develop future policies to ensure consumer protection, and therefore, the debate over the re-evaluation of the use of CO as a protective gas in meat packaging within the EU may be justified. In addition, the application of CO in the EU could be very profitable from the point of view of exports, retail sales in supermarkets, as well as meet consumer quality demands as a value-added technology. In this context, CO-MAP technology has pronounced efficacy to transform the academic research output to industrial application.

9. Conclusions

CO can be utilized at different stages of the value chain of fresh meat, in animal slaughter, distribution, pre-treatment before storage, processing and display packaging. The use of CO in fresh meat packaging gives promising results due to its positive effects on overall meat quality; red color is enhanced and lipid oxidation is reduced. In realistic concentrations, CO as such has no antimicrobial effect, and CO_2 in sufficient concentrations is required for delaying growth of spoilage Gram-negative bacteria, which results in shelf-life prolongation during wider distribution of case-ready products.

The use of CO is especially controversial. Some countries approve the application such as the U.S., while others, as is the case of the EU member states, ban it from food processing due to the potential toxic effect. The commercial application of CO in meat packaging was not then considered feasible because of possible environmental hazards for workers. CO has previously been reported to mask meat spoilage, and this was the primary concern raised for the prohibition as this may mislead consumers.

Risk of CO toxicity from the packaging process or from consumption of CO-treated meats is negligible. Moreover, the addition of CO pre-treatments prior to VP may be beneficial to allow a desirable color to be induced while allowing aging to occur within the package and increasing meat tenderness. Additionally, CO is not present in the pack during storage.

Countries differ not only with respect to regulations, but also with regard to consumers' attitudes towards new technologies. Recent European consumer acceptance studies demonstrate the promising future potential of the application of CO within the EU. Consequently, to improve consumer attitudes about CO packaging of fresh meat, communications among all segments of the meat cold chain should be designed to not only inform consumers about the use of CO, but also familiarize consumers about the science of this technology. In addition, the application of CO in the EU could be very profitable from the point of view of exports, retail sales in supermarkets, as well as meeting consumer quality demands as a value-added technology. Possible problems can arise related to detecting CO treatment of foods. This is a key issue to assure fair competition and to gain consumer's trust. In light of all the data cited in this article, it can be said that the application of CO for the packaging of fresh meat could be re-evaluated. In this context, CO-MAP technology has pronounced efficacy to transform the academic research output to the meat company's application.

Recommendations and future prospects the addressed food industries, consumers and regulators on what would be a "best practice" in the use of CO in food packaging:

- The main disadvantage of using CO for meat packaging is the concern about masking of the microbiological risk by the formation of the stable and longer, bright red color.
- A combination of low levels of CO with CO_2 in high concentrations, which reduce the growth of microorganisms, is of key importance to justify approval by EU regulatory agencies.
- It is also imperative that CO risk management be implemented in the packaging operations. The supermarkets and consumers at home must handle meat with strict hygienic standards, and low storage temperatures must be kept in a continuous chill chain.
- CO followed by vacuum packaging is promising because of the possibility of better adjusting the color stability of the meat to the time of spoilage.
- By adopting CO for meat and fish MAP, the meat industry must label the packages with reliable times for maximum shelf-life (the use-by date).

- The EU considers packaging gases as additives, and CO is not on the list of such gases. In order for CO to be approved as an additive within the EU, the following criteria must be met according to Directive 94/34/EEC. "They present no hazard to the heath of consumer at the level of use proposed, so far as can be judged on the scientific evidence available; to provide aids in manufacture, processing, preparation, treatment, packing, transport or storage of food, provided that the additive is not used to disguise the effects of the use of faulty raw materials or of undesirable (including unhygienic) practices or techniques during the course of any of these activities and to assess the possible harmful effects by evaluation any cumulative, synergistic or potentiating effect of its use and the phenomenon of human intolerance to substances foreign to the body".
- In the EU, current labeling regulations require packages with meat and meat products in MAP to be labeled with "Packaged in a protective atmosphere". The specific gases do not need to be declared on the packages. Conventional gases will probably be the ones most concerned by this statement. Low concentrations of CO in food packaging systems could be required to appear on the product label to inform consumers of its use in the product.
- Analysis of CO can be used to control whether muscle food products have been treated with CO or not, despite not having been labeled as such. These considerations underline the suitability to develop better alternative fast methods to detect even small amounts of the CO-Mb adduct in muscle tissues, in regard to the fraudulent treatment of meat and fish in the MAP systems.

All this promotes high ethical standards in commercial communications by means of effective regulation, for the benefit of consumers and business in the world (Europe and beyond), and this implies that industrialized countries and members of their regulatory agencies must develop coherent and robust systems of regulation that can respond effectively to new challenges.

Acknowledgments: Djamel Djenane thanks the Agencia Española de Cooperación Internacional para el Desarrollo (AECID) for its financial support during his stay at the University of Zaragoza (Spain) as part of his Ph.D thesis preparation. Thanks to Pedro Roncalés and José Antonio Beltràn (University of Zaragoza, Spain) for giving him the great opportunity to be well trained in the field of Meat Science and Technology. Thanks are also addressed to the Algerian Ministry of Higher Education and Scientific Research for its financial support through a grant to "Laboratoire de Qualité et Sécurité des Aliments, LQSA". Djamel Djenane is grateful to the many investigators that have published research in this area.

Author Contributions: Djenane Djamel contributed to: article conception and design, acquisition of data from various published articles, analysis and interpretation of data from published articles and drafting of the manuscript. Roncalés Pedro contributed to: critical revision of the manuscript and gave final approval of the version to be submitted.

Conflicts of Interest: The authors declare no conflicts of interest.

References

1. Renerre, M.; Labadie, J. Fresh meat packaging and meat quality. In Proceedings of the 39th International Congress of Meat Science and Technology, Calgary, AB, Canada, 1–6 August 1993. Session 8.
2. Jakobsen, M.; Bertelsen, G. Color stability and lipid oxidation of fresh beef—Development of a response surface model for predicting the effects of temperature, storage time, and modified atmosphere composition. *Meat Sci.* **2000**, *54*, 49–57. [CrossRef]
3. Jayasingh, P.; Cornforth, D.P.; Brennand, C.P.; Carpenter, C.E.; Whittier, D.R. Sensory evaluation of ground beef stored in high-oxygen modified atmosphere packaging. *J. Food Sci.* **2002**, *67*, 3493–3496. [CrossRef]
4. Lund, M.N.; Lametsch, R.; Hviid, M.S.; Jensen, O.N.; Skibsted, L.H. High-oxygen packaging atmosphere influences protein oxidation and tenderness of porcine *Longissimus dorsi* during chill storage. *Meat Sci.* **2007**, *77*, 295–303. [CrossRef] [PubMed]
5. Lindahl, G.; Lagerstedt, A.; Ertbjerg, P.; Sampels, S.; Lundström, K. Ageing of large cuts of beef loin in vacuum or high oxygen modified atmosphere, effect on shear force, calpain activity, desmin degradation and protein oxidation. *Meat Sci.* **2010**, *85*, 160–166. [CrossRef] [PubMed]

6. Hague, M.A.; Warren, K.E.; Hunt, M.C.; Kropf, D.H.; Kastner, C.L.; Stroda, S.L.; Johnson, D.E. Endpoint temperature, internal cooked color, and expressible juice color relationships in ground beef patties. *J. Food Sci.* **1994**, *59*, 465–470. [CrossRef]
7. Killinger, K.M.; Hunt, M.C.; Campbell, R.E.; Kropf, D.H. Factors affecting premature browning during cooking of store-purchased ground beef. *J. Food Sci.* **2000**, *65*, 585–587. [CrossRef]
8. Warren, K.E.; Hunt, M.C.; Kropf, D.H. Myoglobin oxidative state affects internal cooked color development in ground beef patties. *J. Food Sci.* **1996**, *61*, 513–515. [CrossRef]
9. Lyon, B.G.; Berry, B.W.; Soderberg, D.; Clinch, N. Visual color and doneness indicators and the incidence of premature brown color in beef patties cooked to four end point temperatures. *J. Food Prot.* **2000**, *63*, 1389–1398. [CrossRef] [PubMed]
10. USFDA. *GRAS Notice Number GRN 000143*; United States Food and Drug Administration: Washington, DC, USA, 2004.
11. Cornforth, D.P.; Hunt, M.C. Low-oxygen packaging of fresh meat with carbon monoxide: Meat quality, microbiology, and safety. In *The American Meat Science Association (AMSA) White Paper Series Number 2*; American Meat Science Association: Savoy, IL, USA, 2008.
12. Brewer, M.S.; Wu, S.; Field, R.A.; Ray, B. Carbon monoxide effects on color and microbial counts of vacuum packaged beef steaks in refrigerated storage. *J. Food Qual.* **1994**, *17*, 231–236. [CrossRef]
13. Mancini, R.A.; Hunt, M.C. Current research in meat color. *Meat Sci.* **2005**, *71*, 100–121. [CrossRef] [PubMed]
14. Lanier, T.C.; Carpenter, J.A.; Toledo, R.T.; Regan, J.O. Metmyoglobin reduction in beef systems as affected by aerobic, anaerobic, and carbon monoxide-containing environments. *J. Food Sci.* **1978**, *43*, 1788–1792. [CrossRef]
15. Meischen, H.W.; Huffman, D.L.; Davis, G.W. Branded beef—Product of tomorrow today. In Proceedings of the 40th Reciprocal Meat Conference, Saint Paul, MN, USA, 28 June–1 July 1987; pp. 37–46.
16. Sørheim, O.; Nissen, H.; Nesbakken, T. The storage life of beef and pork packaged in an atmosphere with low carbon monoxide and high carbon dioxide. *Meat Sci.* **1999**, *52*, 157–164. [CrossRef]
17. Grebitus, C.; Jensen, H.H.; Roosen, J. US and German consumer preferences for ground beef packaged under a modified atmosphere—Different regulations, different behaviour? *Food Policy* **2013**, *40*, 109–118. [CrossRef]
18. Djenane, D.; Beltrán, J.A.; Camo, J.; Roncalés, P. Influence of vacuum at different ageing times and subsequent retail display on shelf life of beef cuts packaged with active film under high O_2. *J. Food Sci. Technol.* **2016**, *53*, 4244–4257. [CrossRef] [PubMed]
19. Djenane, D.; Sanchéz, A.; Beltrán, J.A.; Roncalés, P. The shelf life of beef steaks treated with DL-lactic acid and antioxidants and stored under modified atmospheres. *Food Microbiol.* **2003**, *20*, 1–7. [CrossRef]
20. Sanchéz, A.; Djenane, D.; Torrescano, G.; Beltran, J.A.; Roncalés, P. The effects of ascorbic acid, taurine, carnosine and rosemary powder on color and lipid stability of beef patties packaged in modified atmosphere. *Meat Sci.* **2001**, *58*, 421–429. [CrossRef]
21. Djenane, D.; Sanchéz, A.; Beltrán, J.A.; Roncalés, P. Ability of α-tocopherol, taurine and rosemary, in combination with vitamin C, to increase the oxidative stability of beef steaks packaged in modified atmosphere. *Food Chem.* **2002**, *76*, 407–415. [CrossRef]
22. Ledward, D.A. Metmyoglobin formation in beef stored in carbon dioxide enriched and oxygen depleted atmospheres. *J. Food Sci.* **1970**, *35*, 33–37. [CrossRef]
23. Kropf, D.H. Effects of retail display conditions on meat color. *Recipr. Meat Conf. Proc.* **1980**, *33*, 15–32.
24. Zakrys, P.I.; Hogan, S.A.; O'Sullivan, M.G.; Allen, P.; Kerry, J.P. Effects of oxygen concentration on the sensory evaluation and quality indicators of beef muscle packed under modified atmosphere. *Meat Sci.* **2008**, *79*, 648–655. [CrossRef] [PubMed]
25. Faustman, C.; Sun, Q.; Mancini, R.; Suman, S.P. Myoglobin and lipid oxidation interactions: Mechanistic bases and control. *Meat Sci.* **2010**, *86*, 86–94. [CrossRef] [PubMed]
26. Wongwichian, C.; Klomklao, S.; Panpipat, W.; Benjakul, S.; Chaijan, M. Interrelationship between myoglobin and lipid oxidations in oxeye scad (*Selar boops*) muscle during iced storage. *Food Chem.* **2015**, *174*, 279–285. [CrossRef] [PubMed]
27. MacDougall, D.B. Changes in the colour and opacity of meat. *Food Chem.* **1982**, *9*, 75–88. [CrossRef]
28. Djenane, D.; Sanchéz, A.; Beltrán, J.A.; Roncalés, P. Extension of the retail display life of fresh beef packaged in modified atmosphere by varying lighting conditions. *J. Food Sci.* **2001**, *66*, 181–185. [CrossRef]
29. Walker, H.W. Effects of Microflora on Fresh Meat Color. *Recipr. Meat Conference Proc.* **1980**, *33*, 33–36.

30. Bertelsen, G.; Boegh-Soerensen, L. The effect of lighting on color of beef. In Proceedings of the Institutional Investors Roundtable (IIR)–Meeting (Commission C2), Bristol, UK, 10–12 September 1986; pp. 433–438.
31. Andersen, H.J.; Skibsted, L.H. Oxidative stability of frozen pork patties, effect of light and added salt. *J. Food Sci.* **1991**, *56*, 1182–1184. [CrossRef]
32. Sánchez-Escalante, A.; Torrescano, G.; Djenane, D.; Beltrán, J.A.; Giménez, B.; Roncalés, P. Effect of antioxidants and lighting conditions on color and lipid stability of beef patties packaged in high-oxygen modified atmosphere. *CyTA J. Food.* **2011**, *9*, 49–57. [CrossRef]
33. Martínez, L.; Cilla, I.; Beltrán, J.A.; Roncalés, P. Effect of illumination on the display life of fresh pork sausages packaged in modified atmosphere. Influence of the addition of rosemary, ascorbic acid and black pepper. *Meat Sci.* **2007**, *75*, 443–450. [CrossRef] [PubMed]
34. Hood, D.E. Factors affecting the rate of metmyoglobin accumulation in pre-packaged beef. *Meat Sci.* **1980**, *4*, 247–265. [CrossRef]
35. Aurand, L.W.; Boone, N.H.; Gidding, G.G. Superoxide and singlet oxygen in milk lipid peroxidation. *J. Dairy Sci.* **1977**, *60*, 363–369. [CrossRef]
36. Kiritsakis, A.; Dugan, L.R. Studies in photooxidation of olive oil. *J. Am. Oil Chem. Soc.* **1985**, *62*, 892–896. [CrossRef]
37. Luby, J.M.; Gray, J.I.; Harte, B.R. Effects of Packaging and Light Source on the Oxidative Stability of Cholesterol in butter. *J. Food Sci.* **1986**, *51*, 908–911. [CrossRef]
38. Whang, K.; Peng, I.C. Photosensitized Lipid Peroxidation in Ground Pork and Turkey. *J. Food Sci.* **1988**, *53*, 1596–1598. [CrossRef]
39. Lee, C.K.; Li, S.; Hui, S.Y. A design methodology for smart LED lighting systems powered by weakly regulated renewable power grids. *IEEE Trans. Smart Grid* **2011**, *2*, 548–554. [CrossRef]
40. Mahalik, N.P. Advances in Packaging Methods, Processes and Systems. *Challenges* **2014**, *5*, 374–389. [CrossRef]
41. Djenane, D.; Sanchéz, A.; Beltrán, J.A.; Roncalés, P. Extension of the shelf life of beef steaks packaged in a modified atmosphere by treatment with rosemary and display under UV-free lighting. *Meat Sci.* **2003**, *64*, 417–426. [CrossRef]
42. Siegel, D.G. Case-ready concepts: Packaging technologies. In Proceedings of the Western Science Research Update Conference on Technologies for Improving the Quality and Safety of Case-Ready Products, Annual Meeting of the National Meat Association, Las Vegas, NV, USA, 22 February 2001.
43. Gill, C. The solubility of carbon dioxide in meat. *Meat Sci.* **1988**, *22*, 65–71. [CrossRef]
44. Gee, D.L.; Brown, W.D. Extension of shelf life in refrigerated ground beef stored under an atmosphere containing carbon dioxide and carbon monoxide. *J. Agric. Food Chem.* **1978**, *26*, 274–276. [CrossRef] [PubMed]
45. Seideman, S.C.; Durland, P.R. The utilization of modified gas atmosphere packaging for fresh meat: A review. *J. Food Qual.* **1984**, *6*, 239–252. [CrossRef]
46. Wolfe, S.K. Use of CO- and CO_2-enriched atmospheres for meats, fish, and produce. *Food Technol.* **1980**, *34*, 55–58.
47. Jongberg, S.; Wen, J.; Tørngren, M.A.; Lund, M.N. Effect of high-oxygen atmosphere packaging on oxidative stability and sensory quality of two chicken muscles during chill storage. *Food Packag. Shelf Life* **2014**, *1*, 38–48. [CrossRef]
48. Brewer, M.S.; Wu, S.Y. Display, packaging and meat block location effects on color and lipid oxidation of frozen lean ground beef. *J. Food Sci.* **1993**, *58*, 1219–1223. [CrossRef]
49. Li, X.; Lindahl, G.; Zamaratskaia, G.; Lundström, K. Influence of vacuum skin packaging on color stability of beef *Longissimus lumborum* compared with vacuum and high-oxygen modified atmosphere packaging. *Meat Sci.* **2012**, *92*, 604–609. [CrossRef] [PubMed]
50. Lagerstedt, Å.; Ahnström, M.L.; Lundström, K. Vacuum skin pack of beef—A consumer friendly alternative. *Meat Sci.* **2011**, *88*, 391–396. [CrossRef] [PubMed]
51. Lawrence, T.E.; Kropf, D.H. Packaging Vacuum. In *Encyclopedia of Meat Sciences*, 2nd ed.; Academic Press: Cambridge, MA, USA, 2014; pp. 26–33.
52. Djenane, D.; Martinez, L.; Sanchéz, A.; Beltrán, J.A.; Roncalés, P. Antioxidant effect of carnosine and carnitine in fresh beef steaks stored under modified atmosphere. *Food Chem.* **2004**, *85*, 453–459. [CrossRef]

53. Sanchéz-Escalante, A.; Djenane, D.; Torrescano, G.; Beltrán, J.A.; Roncalés, P. Antioxidant action of borage, rosemary, oregano and ascorbic acid in beef patties packaged in modified atmosphere. *J. Food Sci.* **2003**, *68*, 339–344. [CrossRef]
54. Sørheim, O.; Aune, T.; Nesbakken, T. Technological, hygienic and toxicological aspects of carbon monoxide used in modified-atmosphere packaging of meat. *Trends Food Sci. Technol.* **1997**, *8*, 307–312. [CrossRef]
55. Ishiwata, H.; Takeda, Y.; Kawasaki, Y.; Yoshida, R.; Sugita, T.; Sakamoto, S. Concentration of carbon monoxide in commercial fish flesh exposed to carbon monoxide gas for colour fixing. *J. Food Hyg. Soc. Jpn.* **1996**, *37*, 83–90. [CrossRef]
56. Durante, W.; Schafer, A.I. Carbon monoxide and vascular cell function. *Int. J. Mol. Med.* **1998**, *2*, 255–262. [CrossRef] [PubMed]
57. Dauphin, J.F.; Saint Lebe, L.R. Radiation chemistry of carbohydrates. In *Radiation Chemistry of Major Food Components*; Elias, P.S., Cohen, A.J., Eds.; Elsevier: Amsterdam, The Netherlands, 1997; pp. 131–185.
58. Nam, K.C.; Ahn, D.U. Carbon monoxide-heme pigment is responsible for the pink color in irradiated raw turkey breast meat. *Meat Sci.* **2002**, *60*, 25–33. [CrossRef]
59. Nam, K.C.; Ahn, D.U. Mechanisms of pink color formation in irradiated precooked turkey breast meat. *J. Food Sci.* **2002**, *67*, 600–607. [CrossRef]
60. Kim, Y.H.; Nam, K.C.; Ahn, D.U. Color, Oxidation-Reduction Potential, and Gas Production of Irradiated Meats from Different Animal Species. *J. Food Sci.* **2002**, *67*, 1692–1695. [CrossRef]
61. Ismail, H.A.; Lee, E.J.; Ko, K.Y.; Paik, H.D.; Ahn, D.U. Effect of Antioxidant Application Methods on the Color, Lipid Oxidation, and Volatiles of Irradiated Ground Beef. *J. Food Sci.* **2009**, *74*, 25–32. [CrossRef] [PubMed]
62. Lee, E.J.; Ahn, D.U. Sources and Mechanisms of Carbon Monoxide Production by Irradiation. *J. Food Sci.* **2004**, *69*, C485–C490. [CrossRef]
63. Lee, E.J.; Ahn, D.U. Sources and mechanisms of carbon monoxide production by irradiation. In Proceedings of the Annual meeting of the Institute of Food Technologists, Chicago, IL, USA, 16–20 June 2003. Session 76E.
64. Furuta, M.; Dohmaru, T.; Katayama, T.; Torantoni, H.; Takeda, A. Detection of irradiated frozen meat and poultry using carbon monoxide gas as a probe. *J. Agric. Food Chem.* **1992**, *40*, 1099–1100. [CrossRef]
65. Haldane, J. The relation of the action of carbonic oxide to oxygen tension. *J. Physiol.* **1895**, *18*, 201–217. [CrossRef] [PubMed]
66. Lambooy, E.; Spanjaard, W. Euthanasia of young pigs with carbon monoxide. *Vet. Rec.* **1980**, *107*, 59–61. [CrossRef] [PubMed]
67. Kao, L.W.; Nañagas, K.A. Carbon monoxide poisoning. *Med. Clin. N. Am.* **2005**, *89*, 1161–1194. [CrossRef] [PubMed]
68. Kao, L.W.; Nañagas, K.A. Carbon monoxide poisoning. *Emerg. Med. Clin. N. Am.* **2004**, *22*, 985–1018. [CrossRef] [PubMed]
69. Jung, C.R.; Lin, Y.T.; Hwang, B.F. Air pollution and newly diagnostic autism spectrum disorders: A population-based cohort study in Taiwan. *PLoS ONE* **2013**, *8*, e75510. [CrossRef] [PubMed]
70. Frydman, M. The smoking addiction of pregnant women and the consequences on their offspring's intellectual development. *J. Environ. Pathol. Toxicol. Oncol.* **1996**, *15*, 169–172. [PubMed]
71. Brunori, M.; Vallone, B. A globin for the brain. *FASEB J.* **2006**, *20*, 2192–2197. [CrossRef] [PubMed]
72. Barinaga, M. Carbon monoxide: Killer to brain messenger in one step. *Science* **1993**, *15*, 259–309. [CrossRef]
73. Verma, A.; Hirsch, D.J.; Glatt, C.E.; Ronnett, G.V.; Snyder, S.H. Carbon monoxide: A putative neural messenger. *Science* **1993**, *259*, 381–384. [CrossRef] [PubMed]
74. Foresti, R.; Bani-Hani, M.G.; Motterlini, R. Use of carbon monoxide as a therapeutic agent: Promises and challenges. *Intensiv. Care Med.* **2008**, *34*, 649–658. [CrossRef] [PubMed]
75. Otterbein, L.E.; Zuckerbraun, B.S.; Haga, M.; Liu, F.; Song, R.; Usheva, A.; Stachulak, C.; Bodyak, N.; Smith, R.N.; Csizmadia, E.; et al. Carbon monoxide suppresses arteriosclerotic lesions associated with chronic graft rejection and with balloon injury. *Nat. Med.* **2003**, *9*, 183–190. [CrossRef] [PubMed]
76. Cheng, Y.; Mitchell-Flack, M.J.; Wang, A.; Levy, R.J. Carbon monoxide modulates cytochrome oxidase activity and oxidative stress in the developing murine brain during isoflurane exposure. *Free Radic. Biol. Med.* **2015**, *86*, 191–199. [CrossRef] [PubMed]

77. Bauer, I.; Pannen, B.H. Carbon monoxide—From mitochondrial poisoning to therapeutic use. *Crit. Care* **2009**, *13*, 220–229. [CrossRef] [PubMed]
78. Li, L.; Hsu, A.; Moore, P.K. Actions and interactions of nitric oxide, carbon monoxide and hydrogen sulphide in the cardiovascular system and in inflammation—A tale of three gases! *Pharmacol. Ther.* **2009**, *123*, 386–400. [CrossRef] [PubMed]
79. Soni, H.; Pandya, G.; Patel, P.; Acharya, A.; Jain, M.; Mehta, A.A. Beneficial effects of carbon monoxide-releasing molecule-2 (CORM-2) on acute doxorubicin cardiotoxicity in mice: Role of oxidative stress and apoptosis. *Toxicol. Appl. Pharmacol.* **2011**, *253*, 70–80. [CrossRef] [PubMed]
80. Roderique, J.D.; Josef, C.S.; Feldman, M.J.; Spiess, B.D. A modern literature review of carbon monoxide poisoning theories, therapies, and potential targets for therapy advancement. *Rev. Toxicol.* **2015**, *334*, 45–58. [CrossRef] [PubMed]
81. Kim, H.J.; Joe, Y.; Yu, J.K.; Chen, Y.; Jeong, S.O.; Mani, N.; Cho, G.J.; Pae, H.O.; Ryter, S.W.; Chung, H.T. Carbon monoxide protects against hepatic ischemia/reperfusion injury by modulating the miR-34a/SIRT1 pathway. *Biochim. Biophys. Acta* **2015**, *1852*, 1550–1559. [CrossRef] [PubMed]
82. Onyiah, J.C.; Sheikh, S.Z.; Maharshak, N.; Steinbach, E.C.; Russo, S.M.; Kobayashi, T.; Mackey, L.C.; Hansen, J.J.; Moeser, A.J.; Rawls, J.F.; et al. Carbon monoxide and heme oxygenase-1 prevent intestinal inflammation in mice by promoting bacterial clearance. *Gastroenterology* **2013**, *144*, 789–798. [CrossRef] [PubMed]
83. Steiger, C.; Lühmann, T.; Meinel, L. Oral drug delivery of therapeutic gases-Carbon monoxide release for gastrointestinal diseases. *J. Control. Release* **2014**, *189*, 46–53. [CrossRef] [PubMed]
84. Wilkinson, B.H.P.; Janz, J.A.M.; Morel, P.C.H.; Purchas, R.W.; Hendriks, W.H. The effect of modified atmosphere packaging with carbon monoxide on the storage quality of master-packaged fresh pork. *Meat Sci.* **2006**, *73*, 605–610. [CrossRef] [PubMed]
85. Sørheim, O.; Nissen, H.; Aune, T.; Nesbakken, T. Use of carbon monoxide in retail meat packaging. In Proceedings of the 54th Reciprocal Meat Conference, Indianapolis, IN, USA, 24–28 July 2001; pp. 47–51.
86. De Santos, F.; Rojas, M.; Lockhorn, G.; Brewer, M.S. Effect of carbon monoxide in modified atmosphere packaging, storage time and endpoint cooking temperature on the internal color of enhanced pork. *Meat Sci.* **2007**, *77*, 520–528. [CrossRef] [PubMed]
87. El-Badawi, A.A.; Cain, R.F.; Samuels, C.E. Anglemeier, A.F. Color and pigment stability of packaged refrigerated beef. *Food Technol.* **1964**, *18*, 159–163.
88. Luño, M.; Roncalés, P.; Djenane, D.; Beltrán, J.A. Beef shelf life in low O_2 and high CO_2 atmospheres containing different low CO concentrations. *Meat Sci.* **2000**, *55*, 413–419. [CrossRef]
89. Nissen, H.; Alvseike, O.; Bredholt, S.; Hoick, A.; Nesbakken, T. Comparison between the growth of *Yersinia enterocolitica, Listeria monocytogenes, Escherichia coli* O157:H7 and *Salmonella* spp. in ground beef packaged by three commercially used packaging techniques. *Int. J. Food Microbiol.* **2000**, *59*, 211–220. [CrossRef]
90. Jayasingh, P.; Cornforth, D.P.; Carpenter, C.E.; Whittier, D. Evaluation of carbon monoxide treatment in modified atmosphere packaging or vacuum packaging to increase color stability of fresh beef. *Meat Sci.* **2001**, *59*, 317–324. [CrossRef]
91. Kusmider, E.A.; Sebranek, J.G.; Lonergan, S.M.; Honeyman, M.S. Effects of carbon monoxide packaging on color and lipid stability of irradiated ground beef. *J. Food Sci.* **2002**, *67*, 3463–3468. [CrossRef]
92. Krause, T.R.; Sebranek, J.G.; Rust, R.E.; Honeyman, M.S. Use of carbon monoxide packaging for improving the shelf life of pork. *J. Food Sci.* **2003**, *68*, 2596–2603. [CrossRef]
93. John, L.; Cornforth, D.P.; Carpenter, C.E.; Sørheim, O.; Pettee, B.; Whittier, D.R. Comparison of color and thiobarbituric acid (TBA) values of cooked hamburger patties after storage of fresh beef chubs in modified atmospheres. *J. Food Sci.* **2004**, *69*, 608–614. [CrossRef]
94. John, L.; Cornforth, J.; Carpenter, C.E.; Sørheim, O.; Pettee, B.C.; Whittier, D.R. Color and thiobarbituric acid values of cooked top sirloin steaks packaged in modified atmospheres of 80% oxygen, or 0.4% carbon monoxide, or vacuum. *Meat Sci.* **2005**, *69*, 441–449. [CrossRef] [PubMed]
95. Martínez, L.; Djenane, D.; Cilla, I.; Beltrán, J.A.; Roncalés, P. Effect of different concentrations of carbon dioxide and low concentration of carbon monoxide on the shelf-life of fresh pork sausages packaged in modified atmosphere. *Meat Sci.* **2005**, *71*, 563–570. [CrossRef] [PubMed]

96. Wicklund, R.A.; Paulson, D.D.; Tucker, E.M.; Stetzer, A.J.; De Santos, F.; Rojas, M.; MacFarlane, B.J.; Brewer, M.S. Effect of carbon monoxide and high oxygen modified atmosphere packaging and phosphate enhanced, case-ready pork chops. *Meat Sci.* **2006**, *74*, 704–709. [CrossRef] [PubMed]
97. Sørheim, O.; Langsrud, Ø.; Cornforth, D.P.; Johannessen, T.C.; Slinde, E.; Berg, P.; Nesbakken, T. Carbon Monoxide as a Colorant in Cooked or Fermented Sausages. *J. Food Sci.* **2006**, *71*, 549–555. [CrossRef]
98. Stetzer, A.J.; Wicklund, R.A.; Paulson, D.D.; Tucker, E.M.; Macfarlane, B.J.; Brewer, M.S. Effect of carbon monoxide and high oxygen modified atmosphere packaging (MAP) on quality characteristics of beef strip steaks. *J. Muscle Foods* **2007**, *18*, 56–66. [CrossRef]
99. Aspé, E.; Roeckel, M.; Martí, M.C.; Jiménez, R. Effect of pre-treatment with carbon monoxide and film properties on the quality of vacuum packaging of beef chops. *Packag. Technol. Sci.* **2008**, *21*, 395–404. [CrossRef]
100. Mantilla, D.; Kristinsson, H.G.; Balaban, M.O.; Otwell, W.S.; Chapman, F.A.; Raghavan, S. Color stability of frozen whole tilapia exposed to pre-mortem treatment with carbon monoxide. *J. Sci. Food Agric.* **2008**, *88*, 1394–1399. [CrossRef]
101. Linares, M.B.; Bórnez, R.; Vergara, H. Effect of stunning systems on meat quality of Manchego suckling lamb packed under modified atmospheres. *Meat Sci.* **2008**, *78*, 279–287. [CrossRef] [PubMed]
102. Linares, M.B.; Vergara, H. Effect of gas stunning and modified-atmosphere packaging on the quality of meat from Spanish Manchego light lamb. *Small Rumin. Res.* **2012**, *108*, 87–94. [CrossRef]
103. Grobbel, J.P.; Dikemen, M.E.; Hunt, M.C.; Milliken, G.A. Effects of different packaging atmospheres and injection–enhancement on beef tenderness, sensory attributes, desmin degradation, and display color. *J. Anim. Sci.* **2008**, *86*, 2697–2710. [CrossRef] [PubMed]
104. Ramamoorthi, L.; Toshkov, S.; Brewer, M.S. Effects of carbon monoxide-modified atmosphere packaging and irradiation on *E. coli* K12 survival and raw beef quality. *Meat Sci.* **2009**, *83*, 358–365. [CrossRef] [PubMed]
105. Mancini, R.A.; Suman, S.P.; Konda, M.K.R.; Ramanathan, R. Effect of carbon monoxide packaging and lactate enhancement on the color stability of beef steaks stored at 1 °C for 9 days. *Meat Sci.* **2009**, *81*, 71–76. [CrossRef] [PubMed]
106. Fontes, P.R.; Gomide, L.A.M.; Fontes, E.A.F.; Ramos, E.M.; Ramos, A.L.S. Composition and color stability of carbon monoxide treated dried porcine blood. *Meat Sci.* **2010**, *85*, 472–480. [CrossRef] [PubMed]
107. Raines, C.R.; Hunt, M.C. Headspace Volume and Percentage of Carbon Monoxide Affects Carboxymyoglobin Layer Development of Modified Atmosphere Packaged Beef Steaks. *J. Food Sci.* **2010**, *75*, 62–65. [CrossRef] [PubMed]
108. Jeong, J.Y.; Claus, J.R. Color stability of ground beef packaged in a low carbon monoxide atmosphere or vacuum. *Meat Sci.* **2011**, *87*, 1–6. [CrossRef] [PubMed]
109. Bjørlykke, G.A.; Roth, B.; Sørheim, O.; Kvammeb, B.O.; Slinde, E. Effects of carbon monoxide on Atlantic salmon (*Salmo salar* L.). *Food Chem.* **2011**, *127*, 1706–1711. [CrossRef]
110. Suman, S.P.; Mancini, R.A.; Joseph, P.; Ramanathan, R.; Konda, M.K.R.; Dady, G.; Yinb, S. Chitosan inhibits premature browning in ground beef. *Meat Sci.* **2011**, *88*, 512–516. [CrossRef] [PubMed]
111. Ramamoorthi, L.; Toshkov, S.; Brewer, M.S. Effects of irradiation on color and sensory characteristics of carbon monoxide-modified atmosphere packaged beef. *J. Food Process. Preserv.* **2011**, *35*, 701–707. [CrossRef]
112. Pivarnik, LF.; Faustman, C.; Rossi, S.; Suman, S.P.; Palmer, C.; Richard, N.L.; Ellis, P.C.; DiLiberti, M. Quality Assessment of Filtered Smoked Yellowfin Tuna (*Thunnus albacares*) Steaks. *J. Food Sci.* **2011**, *76*, S369–S379. [CrossRef] [PubMed]
113. Venturini, A.C.; Faria, J.A.F.; Olinda, R.A.; Contreras-Castillo, C.J. Shelf Life of Fresh Beef Stored in Master Packages with Carbon Monoxide and High Levels of Carbon Dioxide. *Packag. Technol. Sci.* **2014**, *27*, 29–35. [CrossRef]
114. Lavieri, N.; Williams, S.K. Effects of packaging systems and fat concentrations on microbiology, sensory and physical properties of ground beef stored at 4 ± 1 °C for 25 days. *Meat Sci.* **2014**, *97*, 534–541. [CrossRef] [PubMed]
115. Liu, C.; Zhang, Y.; Yang, X.; Liang, R.; Mao, Y.; Hou, X.; Lu, X.; Luo, X. Potential mechanisms of carbon monoxide and high oxygen packaging in maintaining color stability of different bovine muscles. *Meat Sci.* **2014**, *97*, 189–196. [CrossRef] [PubMed]

116. Concollato, A.; Parisi, G.; Olsen, R.E.; Kvamme, B.O.; Slinde, E.; Dalle Zotte, A. Effect of carbon monoxide for Atlantic salmon (*Salmo salar* L.) slaughtering on stress response and fillet shelf life. *Aquaculture* **2014**, *433*, 13–18. [CrossRef]
117. Rogers, H.B.; Brooks, J.C.; Martin, J.N.; Tittor, A.; Miller, M.F.; Brashears, M.M. The impact of packaging system and temperature abuse on the shelf life characteristics of ground beef. *Meat Sci.* **2014**, *97*, 1–10. [CrossRef] [PubMed]
118. Pereira, A.D.; Gomide, L.A.M.; Cecon, P.R.; Fontes, E.A.F.; Fontes, P.R.; Ramos, E.M.; Vidigal, J.G. Evaluation of mortadella formulated with carbon monoxide-treated porcine blood. *Meat Sci.* **2014**, *97*, 164–173. [CrossRef] [PubMed]
119. Fontes, P.R.; Gomide, L.A.M.; Costa, N.M.B.; Peternelli, L.A.; Fontes, E.A.F.; Ramos, E.M. Chemical composition and protein quality of mortadella formulated with carbon monoxide-treated porcine blood. *LWT-Food Sci. Technol.* **2015**, *64*, 926–931. [CrossRef]
120. Yang, X.; Zhang, Y.; Zhu, L.; Han, M.; Gao, S.; Luo, X. Effect of packaging atmospheres on storage quality characteristics of heavily marbled beef longissimus steaks. *Meat Sci.* **2016**, *117*, 50–56. [CrossRef] [PubMed]
121. Sakowska, A.; Guzek, D.; Głąbska, D.; Wierzbicka, A. Carbon monoxide concentration and exposure time effects on the depth of CO penetration and surface color of raw and cooked beef *Longissimus lumborum* steaks. *Meat Sci.* **2016**, *121*, 182–188. [CrossRef] [PubMed]
122. Lyu, F.; Shen, K.; Ding, Y.; Ma, X. Effect of pretreatment with carbon monoxide and ozone on the quality of vacuum packaged beef meats. *Meat Sci.* **2016**, *117*, 137–146. [CrossRef] [PubMed]
123. Van Rooyen, L.A.; Allen, P.; Crawley, S.M.; O'Connor, D.I. The effect of carbon monoxide pretreatment exposure time on the colour stability and quality attributes of vacuum packaged beef steaks. *Meat Sci.* **2017**, *129*, 74–80. [CrossRef] [PubMed]
124. Sakowska, A.; Guzek, D.; Sun, D.-W.; Wierzbicka, A. Effects of 0.5% carbon monoxide in modified atmosphere packagings on selected quality attributes of *M. Longissimus dorsi* beef steaks. *J. Food Proc. Eng.* **2017**, *40*, 12517–12527. [CrossRef]
125. Van Rooyen, L.A.; Allen, P.; O'Connor, D.I. The application of carbon monoxide in meat packaging needs to be re-evaluated within the EU: An overview. *Meat Sci.* **2017**, *132*, 179–188. [CrossRef] [PubMed]
126. Martínez, L.; Djenane, D.; Cilla, I.; Beltrán, J.A.; Roncalés, P. Effect of varying oxygen concentrations on the shelf-life of fresh pork sausages packaged in modified atmosphere. *Food Chem.* **2006**, *94*, 219–225. [CrossRef]
127. Sanchéz, A.; Torrescano, G.; Djenane, D.; Beltrán, J.A.; Roncalés, P. Stabilisation of colour and odour of beef patties by using lycopene-rich tomato and peppers as a source of antioxidants. *J. Sci. Food Agric.* **2003**, *83*, 187–194. [CrossRef]
128. Ledward, D.A. Post-slaughter influences on the formation of metmyoglobin in beef muscles. *Meat Sci.* **1985**, *15*, 149–171. [CrossRef]
129. Woodruff, R.; Silliker, J. Process and Composition for Producing and Maintaining Good Color in Fresh Meat, Fresh Poultry and Fresh Fish. U.S. Patent No. 4,522,835, 11 June 1985. Available online: http://www.google.com/patents?vid=USPAT4522835 (accessed on 23 January 2008).
130. Clydesdale, F.M.; Francis, F.J. The chemistry of meat color. *Food Prod. Dev.* **1971**, *15*, 581–584.
131. Grant, M.; Clay, B. Accidental carbon monoxide poisoning with severe cardiorespiratory compromise in 2 children. *Am. J. Crit. Care* **2002**, *11*, 128–131. [PubMed]
132. Seyfert, M.; Mancini, R.A.; Hunt, M.C.; Tang, J.; Faustman, C. Influence of carbon monoxide in package atmospheres containing oxygen on colour, reducing activity, and oxygen consumption of five bovine muscles. *Meat Sci.* **2007**, *75*, 432–442. [CrossRef] [PubMed]
133. Jeong, J.Y.; Claus, J.R. Color stability and reversion in carbon monoxide packaged ground beef. *Meat Sci.* **2010**, *85*, 525–530. [CrossRef] [PubMed]
134. Lentz, C.P. Effect of Light Intensity and Other Factors on the Color of Frozen Prepackaged Beef. *Can. Inst. Food Sci. Technol. J.* **1979**, *12*, 47–50. [CrossRef]
135. Tricker, A.R.; Preussmann, R. Carcinogenic N-nitrosamines in the diet: Occurrence, formation, mechanisms and carcinogenic potential. *Mutat. Res.* **1991**, *259*, 277–289. [CrossRef]
136. Bekhit, A.E.D.; Faustman, C. Metmyoglobin reducing activity: Review. *Meat Sci.* **2005**, *71*, 407–439. [CrossRef] [PubMed]

137. Sammel, L.M.; Hunt, M.C.; Kropf, D.H.; Hachmeister, K.A.; Johnson, D.E. Comparison of Assays for Metmyoglobin Reducing Ability in Beef Inside and Outside Semimembranosus Muscle. *J. Food Sci.* **2002**, *67*, 978–984. [CrossRef]
138. King, D.A.; Shackelford, S.D.; Rodriguez, A.B.; Wheeler, T.L. Effect of time of measurement on the relationship between metmyoglobin reducing activity and oxygen consumption to instrumental measures of beef longissimus color stability. *Meat Sci.* **2011**, *87*, 26–32. [CrossRef] [PubMed]
139. Laury, A.; Sebranek, J.G. Use of Carbon Monoxide Combined with Carbon Dioxide for Modified Atmosphere Packaging of Pre- and Postrigor Fresh Pork Sausage To Improve Shelf Life. *J. Food Prot.* **2007**, *70*, 937–942. [CrossRef] [PubMed]
140. Arvanitoyannis, I.S.; Stratakos, A.C. Application of Modified Atmosphere Packaging and Active/Smart Technologies to Red Meat and Poultry: A Review (Review). *Food Bioprocess Technol.* **2012**, *5*, 1423–1446. [CrossRef]
141. Kudra, L.L.; Sebranek, J.G.; Dickson, J.S.; Mendonca, A.F.; Zhang, Q.; Jackson-Davis, A.; Prusa, K.J. Control of *Campylobacter jejuni* in Chicken Breast Meat by Irradiation Combined with Modified Atmosphere Packaging Including Carbon Monoxide. *J. Food Prot.* **2012**, *75*, 1728–1733. [CrossRef] [PubMed]
142. Luño, M.; Beltrán, J.A.; Roncalès, P. Shelf-life extension and colour stabilisation of beef packaged in a low O_2 atmosphere containing CO: Loin steaks and ground meat. *Meat Sci.* **1998**, *48*, 75–84. [CrossRef]
143. Gee, D.L.; Brown, W.D. The effect of carbon monoxide on bacterial growth. *Meat Sci.* **1981**, *5*, 215–222. [CrossRef]
144. Clark, D.S.; Lentz, C.P.; Roth, L.A. Use of carbon monoxide for extending shelf-life of prepackaged fresh beef. *Can. Inst. Food Sci. Technol. J.* **1976**, *9*, 114–117. [CrossRef]
145. Viana, E.; Comide, L.; Vanetti, M. Effect of modified atmospheres on microbiological, color and sensory properties of refrigerated pork. *Meat Sci.* **2005**, *71*, 696–705. [CrossRef] [PubMed]
146. Hunt, M.C.; Mancini, R.A.; Hachmeister, K.A.; Kropf, D.H.; Merriman, M.; DelDuca, G.; Milliken, G. Carbon monoxide in modified atmosphere packaging affects color, shelf life, and microorganisms of beef steaks and ground beef. *J. Food Sci.* **2004**, *69*, C45–C52. [CrossRef]
147. Bórnez, R.; Linares, M.B.; Vergara, H. Microbial quality and lipid oxidation of Manchega breed suckling lamb meat: Effect of stunning method and modified atmosphere packaging. *Meat Sci.* **2009**, *83*, 383–389. [CrossRef] [PubMed]
148. Bórnez, R.; Linares, M.B.; Vergara, H. Effect of different gas stunning methods on Manchega suckling lamb meat packed under different modified atmospheres. *Meat Sci.* **2010**, *84*, 727–734. [CrossRef] [PubMed]
149. Kim, Y.H.; Huff-Lonergan, E.; Sebranek, J.G.; Lonergan, S.M. High-oxygen modified atmosphere packaging system induces lipid and myoglobin oxidation and protein polymerization. *Meat Sci.* **2010**, *85*, 759–767. [CrossRef] [PubMed]
150. Brewer, S. Irradiation effects on meat color—A review. *Meat Sci.* **2004**, *68*, 1–17. [CrossRef] [PubMed]
151. Brewer, M.S. Irradiation effects on meat flavor: A review. *Meat Sci.* **2009**, *81*, 1–14. [CrossRef] [PubMed]
152. Shenoy, K.; Murano, E.A.; Olson, D.G. Survival of heat-shocked Yersinia enterocolitica after irradiation in ground pork. *Int. J. Food Microbiol.* **1998**, *39*, 133–137. [CrossRef]
153. Seyfert, M.; Mancini, R.A.; Hunt, M.C. Internal premature browning in cooked steaks from enhanced beef round muscles packaged in high oxygen and ultra-low oxygen modified atmospheres. *J. Food Sci.* **2004**, *69*, 721–725. [CrossRef]
154. Mermelstein, N.H. Success or failure? How the food technology industrial achievement award winners have fared over the years. *Food Technol.* **1977**, *31*, 36–58.
155. Kader, A.A. Physiological and biochemical effects of carbon monoxide added to controlled atmospheres on fruits. *Acta Hortic.* **1983**, *138*, 221–225. [CrossRef]
156. Mullan, M.; McDowell, D. Modified atmosphere packaging. In *Food Packaging Technology*; Coles, R., McDowell, D., Kirwan, M., Eds.; Blackwell Publishing: London, UK, 2003.
157. Kader, A.A. Advances in CA/MA Applications. Available online: http://ucanr.edu/datastoreFiles/234-105.pdf (accessed on 23 November 2000).
158. Cantwell, M. *Postharvest Handling Systems: Minimally Processed Fruits and Vegetables*; University of California, Cooperative Extension Vegetable Research and Information Center: Oakland, CA, USA, 2005; Available online: http://vric.ucdavis.edu/selectnewtopic.minproc.htm (accessed on 2 December 2007).

159. Zhang, S.; Zhu, L.; Dong, X. Combined Treatment of Carbon Monoxide and Chitosan Reduced Peach Fruit Browning and Softening During Cold Storage. *Int. J. Nutr. Food Sci.* **2015**, *4*, 477–482.
160. Zhang, S.; Yu, Y.; Xiao, C.; Wang, X.; Tian, Y. Effect of carbon monoxide on browning of fresh-cut lotus root slice in relation to phenolic metabolism. *LWT-Food Sci. Technol.* **2013**, *53*, 555–559. [CrossRef]
161. Feng Wang, D.S.; Jayas, N.D.G.; White, P.F. Combined effect of carbon monoxide mixed with carbon dioxide in air on the mortality of stored-grain insects. *J. Stored Prod. Res.* **2009**, *45*, 247–253. [CrossRef]
162. Anderson, C.R.; Wu, W.-H. Analysis of carbon monoxide in commercially treated tuna (*Thunnus* spp.) and mahi–mahi (*Coryphaena hippurus*) by gas chromatography/mass spectrometry. *J. Agric. Food Chem.* **2005**, *53*, 7019–7023. [CrossRef] [PubMed]
163. Kowalski, B. Tasteless Smoke Sources, Specifications, and Controls. In *Modified Atmospheric Processing and Packaging of Fish: Filtered Smokes, Carbon Monoxide, and Reduced Oxygen Packaging*; Otwell, W.S., Kristinsson, H.G., Balaban, M.O., Eds.; Blackwell Publishing Professional: Ames, IA, USA, 2006.
164. Kristinsson, H.; Balaban, M.; Otwell, W.S. Microbial and quality consequences of aquatic foods treated with carbon monoxide or filtered wood smoke. In *Modified Atmospheric Processing and Packaging of Fish; Filtered Smokes, Carbon Monoxide Reduced Oxygen Packaging*; Otwell, W.S., Kristinsson, H.G., Balaban, M.O., Eds.; Blackwell: Ames, IA, USA, 2006; pp. 65–86.
165. Kristinsson, H.; Balaban, M.; Otwell, W.S. The influence of carbon monoxide and filtered smoke on fish muscle colour. In *Modified Atmospheric Processing and Packaging of Fish; Filtered Smokes, Carbon Monoxide Reduced Oxygen Packaging*; Otwell, W.S., Kristinsson, H.G., Balaban, M.O., Eds.; Blackwell: Ames, IA, USA, 2006; pp. 29–52.
166. Mantilla, D.; Kristinsson, H.G.; Balaban, M.O.; Otwell, W.S.; Chapman, F.A.; Raghavan, S. Carbon monoxide treatments to impart and retain muscle colour in tilapia fillets. *J. Food Sci.* **2008**, *73*, C390–C399. [CrossRef] [PubMed]
167. Chow, C.-J.; Hsieh, P.-P.; Tsai, M.-L.; Chu, Y.-J. Quality changes during iced and frozen storage of tuna flesh treated with carbon monoxide gas. *J. Food Drug Anal.* **1998**, *6*, 622–623.
168. Fraqueza, M.J.; Ferreira, M.F.; Ouakinin, J.S.; Barreto, A.S. Effect of carbon monoxide and argon in sliced turkey meat under modified atmosphere packaging (preliminary assays). In Proceedings of the 46th International Congress of Meat Science and Technology, Buenos Aires, Argentina, 27 August–1 September 2000; pp. 760–761.
169. Rozbeh, M.; Kalchayanand, N.; Field, R.A.; Johnson, M.C.; Ray, B. The influence of biopreservatives on the bacterial level of refrigerated vacuum packaged beef. *J. Food Saf.* **1993**, *13*, 99–111. [CrossRef]
170. Carpenter, C.E.; Cornforth, D.P.; Whittier, D. Consumer preferences for beef color and packaging did not affect eating satisfaction. *Meat Sci.* **2001**, *57*, 359–363. [CrossRef]
171. Brooks, J.C.; Alvarado, M.; Stephens, T.P.; Kellermeier, J.D.; Tittor, A.W.; Miller, M.F.; Brashears, M.M. Spoilage and safety characteristics of ground beef packaged in traditional and modified atmosphere packages. *J. Food Prot.* **2008**, *71*, 293–301. [CrossRef] [PubMed]
172. Aaslyng, M.D.; Torngren, M.A.; Madsen, N.T. Scandinavian consumer preference for beef steaks packed with or without oxygen. *Meat Sci.* **2010**, *85*, 519–524. [CrossRef] [PubMed]
173. Hunt, M.C.; Sørheim, O.; Slinde, E. Color and heat denaturation of myoglobin forms in ground beef. *J. Food Sci.* **1999**, *64*, 847–851. [CrossRef]
174. Grebitus, C.; Jensen, H.; Roosen, J.; Sebranek, J. Fresh meat packaging: Consumer acceptance of modified atmosphere packaging including carbon monoxide. *J. Food Prot.* **2013**, *76*, 99–107. [CrossRef] [PubMed]
175. Ohtsuki, T.; Kawasaki, Y.; Kubota, H.; Namiki, T.; Iizuka, T.; Shionoya, N.; Yoshii, N.; Ohara, R.; Tanaka, M.; Kobayashi, H.; et al. Improvement of carbon monoxide analysis in fish meat. *J. Food Hyg. Soc. Jpn.* **2011**, *52*, 130–134. [CrossRef]
176. Droghetti, E.; Bartolucci, G.L.; Focardi, C.; Bambagiotti-Alberti, M.; Nocentini, M.; Smulevich, G. Development and validation of a quantitative spectrophotometric method to detect the amount of carbon monoxide in treated tuna fish. *Food Chem.* **2011**, *128*, 1143–1151. [CrossRef]
177. Bartolucci, G.; Droghetti, E.; Focardi, C.; Bambagiotti-Alberti, M.; Nocentini, M.; Smulevich, G. High throughput headspace GC-MS quantitative method to measure the amount of carbon monoxide in treated tuna fish. *J. Mass Spectrom.* **2010**, *45*, 1041–1045. [CrossRef] [PubMed]
178. World Health Organization (WHO). Carbon monoxide. In *Environmental Health Criteria, 213*; World Health Organization: Geneva, Switzerland, 1999.

179. EPA. National Ambient Air Quality Standards (NAAQS). Available online: http://www.epa.gov/air/criteria.html (accessed on 9 December 2007).
180. Church, N. Developments in modified-atmosphere packaging and related technologies. *Trends Food Sci. Technol.* **1994**, *5*, 345–352. [CrossRef]
181. Madani, I.M.; Khalfan, S.; Khalfan, H.; Jidah, J.; Aladin, M.N. Occupational exposure to carbon monoxide during charcoal meat grilling. *Sci. Total Environ.* **1992**, *114*, 141–147. [CrossRef]
182. European Commission (EC). Health & Consumer Protection Directorate-General. Opinion of the Scientific Committee on Food on the Use of Carbon Monoxide as Component of Packaging Gases IN Modified Atmosphere Packaging. 13 December 2001. Available online: http://europa.eu.int/comm/food/fs/sc/scf/index_en.html (accessed on 18 December 2001).
183. Eilert, E.J. New packaging technologies for the 21st century. *Meat Sci.* **2005**, *71*, 122–127. [CrossRef] [PubMed]
184. Watts, D.A.; Wolfe, S.K.; Brown, W.D. Fate of ^{14}C carbon monoxide in cooked or stored ground beef samples. *J. Agric. Food Chem.* **1978**, *26*, 210–214. [CrossRef] [PubMed]
185. Chow, C.-J.; Chu, Y.-J. Effect of heating on residual carbon monoxide content in co-treated tuna and myoglobin. *J. Food Biochem.* **2004**, *28*, 476–487. [CrossRef]
186. Soni, S.; Andhare, V.V. Quantitative Analysis of Carbon Monoxide in Frozen Foods. *Res. J. Recent Sci.* **2015**, *4*, 76–80.
187. EPA. Air Trends—Carbon Monoxide—National Trends in CO Levels. Available online: http://www.epa.gov/airtrends/carbon.htm (accessed on 7 December 2007).
188. Djenane, D. Antioxidant and Antimicrobial Systems for Shelf-Life Extension of Fresh Beef. Ph.D. Thesis, Department of Animal Production and Food Science, Faculty of Veterinary Sciences, Zaragoza University, Zaragoza, Spain, 24 April 2004; p. 275.
189. Australia New Zealand Food Standards Code—Standard 1.3.3—Processing Aids. Available online: http://www.foodstandards.gov.au/code/Documents/1.3.3%20Processing%20aids%20v157.pdf (accessed on 1 March 2016).
190. Sebranek, J.; Hunt, M.C.; Cornforth, D.P.; Brewer, M.S. Perspective—Carbon monoxide packaging of fresh meat. *Food Technol.* **2006**, *60*, 184.
191. Sebranek, J.; Houser, T. Use of CO for red meats: Current research and recent regulatory approvals. In *Modified Atmospheric Processing and Packaging of Fish*; Otwell, W.S., Ed.; Blackwell Publishing: Ames, IA, USA, 2006.
192. USFDA. *GRAS Notice Number GRN 000083*; United States Food and Drug Administration: Washington, DC, USA, 2002.

Article

Volatile Profile of Raw Lamb Meat Stored at 4 ± 1 °C: The Potential of Specific Aldehyde Ratios as Indicators of Lamb Meat Quality

Ioannis Konstantinos Karabagias

Laboratory of Food Chemistry, Department of Chemistry, University of Ioannina, 45110 Ioannina, Greece; ikaraba@cc.uoi.gr; Tel.: +30-697-828-6866

Received: 22 February 2018; Accepted: 14 March 2018; Published: 16 March 2018

Abstract: The objectives of the present study were: (a) to evaluate the aroma evolution of raw lamb packaged in multi-layer coating film and stored at 4 ± 1 °C, with respect to storage time and (b) to investigate whether specific aldehyde ratios could serve as markers of lamb meat freshness and degree of oxidation. Volatile compounds were determined using headspace solid phase microextraction coupled to gas chromatography/mass spectrometry. Results showed that the most dominant volatiles were 2,2,4,6,6-pentamethyl-heptane, hexanal, 1-octen-3-ol, 1-hexanol, carbon disulfide and *p*-cymene. Volatile compound content was increased during storage time. However, statistically significant differences were recorded only for hexanal, heptanal, and nonanal ($p < 0.05$). Additionally, the evolution of aldehydes during storage recorded a positive Pearson's correlation (r) ($p < 0.05$), whereas hexanal to nonanal, heptanal to nonanal, octanal to nonanal ratios, along with the sum of aldehydes to nonanal ratio, were positively correlated (r = 0.83–1.00) with the degree of oxidation (mg malonic dialdehyde per kg of lamb meat). A perfect Pearson's correlation (r = 1) was obtained for the ratio hexanal to nonanal. Therefore, this ratio is proposed as an indicator of lamb meat freshness and overall quality.

Keywords: lamb meat; freshness; volatile compounds; aldehydes; aldehyde ratios; quality control

---

1. Introduction

Colour and flavour are the most important quality attributes consumers use to evaluate meat quality, and thus, acceptability and preference. Raw meat has little odour and only a mild serum-like taste, which is described as salty, metallic and "bloody" with a sweet aroma [1].

Sheep meat is a good source of polyunsaturated (linoleic (C18:2, *n*-6) and α-linolenic (C18:3, *n*-3)) and monounsaturated (oleic acid (C18:1 *n*-9)) fatty acids which have beneficial effects for human health. On the other hand, sheep meat contains significant amounts of saturated fatty acids with a prospective negative impact on human health, when consumed daily in high amounts [2].

Among different factors which influence the meat flavour, animal diet, especially its lipid content, is the most important, as the main source of volatile compounds [3]. The lipid fraction of meat, especially phospholipids, undergoes autoxidation, producing a large number of volatile compounds such as acids, aldehydes, ketones and alcohols and promoting the formation of other components such as nitrogen and sulphur-containing compounds [3]. In addition, lamb age, gender and weight may also affect the volatile profile of lamb meat [2]. What should not be forgotten is that biological processes such as rigor-mortis or post-mortem glycolytic fluxes might modify the volatile fraction of fresh meat [4].

The characteristic flavour of fresh meat, fermented meat products or dry cured ham has been reported to be a subtle balance between non-volatile compounds with taste properties and volatile compounds which interact with each other along with proteins and lipids [1,5].

Many methods have been developed to enhance the efficiency of the isolation of the volatiles in foods, such as vacuum distillation, simultaneous distillation and extraction, purge and trap or headspace techniques. Solid-phase microextraction (SPME) is an inexpensive, easy, rapid and efficient technique as compared to others [6–10].

With respect to volatile compound analysis, gas chromatography coupled to mass spectrometry (GC-MS) is the technique that is most generally used. The literature involving research on volatile compounds in raw meat generally considers beef, duck, pork and poultry meat. However, very few publications deal with the volatile fraction originating from refrigerated storage and the large majority of the works deal with meat products and cooked meat using emerging technologies [1,7,11–13]. Surprisingly, there are only a few studies that have been carried out on the determination of volatile compounds in raw meat by using mild temperature analytical procedures [1,11–15] and/or the use of specific volatile compounds for the differentiation of meat origin [13]. In addition, the specific sense of smell that products of animal origin possess may provide useful information regarding the analysis of quality control [16,17].

Lamb meat is an important source of monetary profit for meat producers, exporters, local butcher's shops and supermarkets in the region of Epirus and holds a special position in the food chain and casualties of local people.

Based on the aforementioned, the aims of the present study were: (a) to evaluate the volatile profile of raw lamb meat during storage under refrigeration and (b) to investigate whether specific aldehyde ratios could be correlated with shelf life and degree of oxidation test data [15], serving thus as markers of lamb meat freshness and overall quality.

2. Materials and Methods

2.1. Lamb Meat Samples, Packaging and Analysis Conditions

Leg of lamb meat (hind shank) samples in chunks, of dimensions ca. $2 \times 2 \times 2$ cm, were provided from different chilled carcasses, 24 h post slaughter, by a local meat processing company (SVEKI S.A., Rodotopi, Ioannina, Greece). Lambs belonged to the Fries–Arta breed, which is a cross breed with high-yielding animals, farmed intensively in the lowlands. Lambs were 3–6 months old, male, with a slaughter weight of 12–13 kg. To avoid any source of contamination, lamb meat samples were transported to the laboratory in insulated polystyrene boxes on ice within 1 h of the chopping process. Prepared samples consisted of 150 ± 10 g in weight, were immediately placed in low density polyethylene/polyamide/low density polyethylene (LDPE/PA/LDPE) barrier pouches (6–7 chunks per pouch), 75 μm in thickness, having an oxygen permeability of 52.2 $cm^3\ m^{-2}\ day^{-1}\ atm^{-1}$, at 75% relative humidity (RH), 23 °C and a water vapour permeability of 2.4 $g\ m^{-2}\ day^{-1}$ at 100% RH, 23 °C. The experimental procedure was carried out on 36 samples: 2 bags $\times$ 2 replicates $\times$ 9 sampling days. Samples were stored at -18 °C in order to avoid any source of deterioration and prepared daily prior to headspace solid phase microextraction coupled to gas chromatography/mass spectrometry analysis (HS-SPME-GC/MS).

2.2. Determination of Lipid Oxidation

The degree of lipid oxidation in lamb meat samples was estimated using thiobarbituric acid assay (TBA) according to the methodology described in a previous work [18]. The value of TBA shows the degree of oxidation of the fats of a product. Measuring the degree of oxidation of polyunsaturated fatty acids is an indicator of the stability of a food's fat in oxidation. In addition, the value of TBA shows the content of malondialdehyde (MDA) (mg/kg) in a fatty foodstuff since the latter compound originates from lipid peroxidation of unsaturated fatty acids. During the reaction, two molecules of TBA and one molecule of MDA react to give a red-coloured product which may be determined spectrophotometrically at 532 nm [18]. Absorbance measurements were accomplished using a UV-VIS spectrophotometer (PerkinElmer, Lambda 25, East Lyne, CT, USA).

2.3. Determination of Volatile Compounds

A divinyl benzene/carboxen/polydimethylsiloxane (DVB/CAR/PDMS) fiber 50/30 μm (Supelco, Bellefonte, PA, USA) was used to extract headspace volatile compounds from raw lamb meat. Prior to use, the fiber was conditioned following the manufacturer's recommendations. Approximately, 2 g of lamb meat from different parts of the lamb legs were placed in a 20 mL crimp-cap vial (72 × 20 mm) equipped with PTFE/silicone septa. Then, 10 μL of an internal standard (4-methyl-2-pentanone, of initial concentration C = 640 ng/mL, Sigma Aldrich (Darmstadt, Germany)) was added at the walls of the vial, and the vial was wrapped. It was vortexed for 5 min in order for the internal standard to be spread homogenously at the surface of raw lamb meat. The vial was then maintained at 50 °C in a water bath under stirring at 600 rpm during the entire extraction procedure, which was 30 min (15 min was the equilibration time and 15 min was the adsorption time). It should be stressed that when the heating of the raw meat product is somehow "mild" during analysis, the volatile profile can be considered as pure information, originating from raw meat. Volatile compounds analysis was carried out on day 1, day 5, and day 9 of storage. Each sample was run in duplicate (n = 2).

2.4. GC/MS Instrumentation and Method Conditions

A Hewlett-Packard, model 6890, gas chromatograph coupled to a Hewlett-Packard, model 5973, mass spectrometric detector (Agilent Technologies, Wilmington, DE, USA) was used for the determination of lamb meat volatile compounds. A HP-INNOWAX (polyethylene glycol capillary column, Agilent, Santa Clara, CA, USA) (30 m × 320 μm i.d., × 0.50 μm film thickness) was used, with helium as the carrier gas (purity 99.999%) at 1.4 mL/min flow rate. The injector and MS-transfer line were maintained both at 250 °C, respectively. For the analysis of lamb meat volatile compounds, oven temperature was held at 40 °C for 5 min, increased to 110 °C at 5 °C/min, and finally further increased to 240 °C at 8 °C/min (5 min hold). The fiber was maintained in the injector for 10 min. Electron impact mass spectra were recorded at 29–300 mass range with 3 scans/s. An electron ionization system was used with ionization energy of 70 eV. Regarding the detector, the temperature of MS quadrapole was held at 150 °C and that of MS source at 230 °C, respectively. Finally, a split ratio 2:1 was used.

2.5. Mass Spectral Data Processing

The identification of compounds was achieved by comparing the mass spectra of the chromatographic peaks with those of the Wiley 275 MS database. For semi-quantification, compound concentration was expressed as ng/g based on the ratio: peak area of analyte/peak area of internal standard (4-methyl-2-pentanone), considering the density (0.8 g/mL) of the internal standard. For validation purposes, volatile compounds having only ≥80% similarity with the Wiley mass spectral library were tentatively identified using GC-MS spectra. For the determination of linear retention indices, a mixture of n-alkanes (C_6–C_{20}) dissolved in n-hexane was employed. The mixture was supplied by Sigma Aldrich (Darmstadt, Germany). The calculation was carried out for components eluting between n-octane and n-eicosane using Equation (1).

2.6. Formatting of Mathematical Components

Kovats indices were calculated using Equation (1):

$$I = 100 \times \left(n + \left(\frac{tr(unkown) - tr(n)}{tr(N) - tr(n)}\right)\right) \quad (1)$$

where I = {\displaystyle I=}, I = Kovats retention index; n = {\displaystyle n=}, n = the number of carbon atoms in the smaller n-alkane; N = {\displaystyle N=}, N = the number of carbon atoms in the larger n-alkane; tr = {\displaystyle t_{r}=}, tr = the retention time.

Equation (1) is the main source for the calculation of Kovats index values, since the temperature was programmed.

2.7. Statistical Analysis

The evolution of each volatile compound with respect to storage time was estimated using paired samples *t*-test analysis at the confidence level $p \leq 0.05$. Correlations were obtained by Pearson's bivariate correlation coefficient (r), at the confidence level $p < 0.01$. All statistical treatments were accomplished using the SPSS version 20.0 statistics (IBM, Armonk, NY, USA) software.

3. Results

Significant variations were observed in the volatile fraction of raw lamb meat during storage under refrigeration, including hexanal, heptanal, and nonanal, as specified by paired samples *t*-test ($p \leq 0.05$). There was observed an increasing trend in all volatile compounds with respect to storage time, with the exception of octanal, toluene, and carbon disulfide, where fluctuations were recorded (Table 1). A typical gas chromatogram, pointing out the volatile compounds identified in the headspace of raw lamb meat is given in Figure 1.

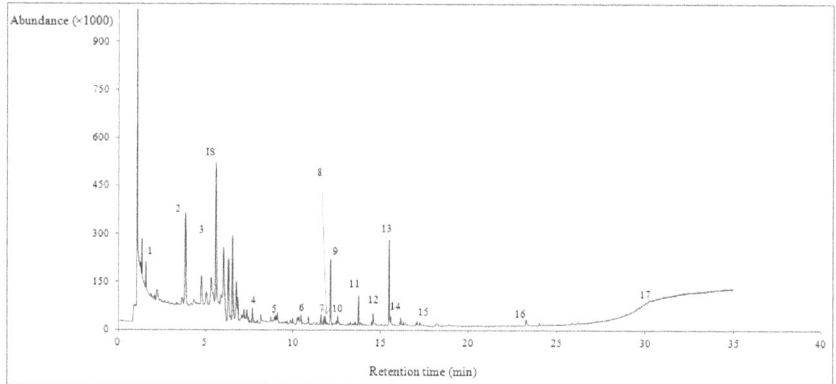

Figure 1. A typical gas chromatogram of raw lamb meat on day 5 of storage. Volatile compounds are numbered according to retention time given in Table 1. 1: carbon disulfide, 2: 2,2,4,6,6-pentamethyl -heptane, 3: toluene, 4: hexanal, 5: *o*-xylene, 6: heptanal, 7: 1-pentanol, 8: 3-octanone, 9: *p*-cymene, 10: octanal, 11: 1-hexanol, 12 :nonanal, 13: 1-octen-3-ol, 14: 1-heptanol, 15: 1-octanol, 16: benzothiazole, 17: 15-crown ether. IS: internal standard.

Table 1. Volatile compounds (average ± SD, ng/g) determined in raw lamb meat under air packaging with respect to storage time (days).

VOCs	RT	KI	Day 1	Day 5	Day 9	MSLM (%)	MI	*p*
			Content (ng/g)	Content (ng/g)	Content (ng/g)			
Alcohols								
1-Pentanol	11.77	1233	11.77 ± 0.01	24.73 ± 0.01	56.77 ± 7.83	83	MS/KI	$p = 0.144$
1-Hexanol	13.77	1342	27.70 ± 0.01	71.29 ± 47.72	236.92 ± 19.62	83	MS/KI	$p = 0.224$
1-Octen-3-ol	15.49	1459	96.45 ± 0.01	145.40 ± 156.77	397.11 ± 25.70	90	MS/KI	$p = 0.151$
1-Heptanol	15.59	1461	11.98 ± 0.01	37.40 ± 4.91	63.20 ± 8.69	86	MS/KI	$p = 0.121$
1-Octanol	17.29	1557	23.55 ± 0.02	30.90 ± 0.01	57.27 ± 16.65	91	MS/KI	$p = 0.057$
Aldehydes								
Hexanal	7.75	1105	163.20 ± 0.01	231.02 ± 92.34	340.37 ± 165.89	96	MS/KI	$p = 0.040$
Heptanal	10.34	1185	36.64 ± 0.02	29.38 ± 0.76	60.94 ± 6.75	98	MS/KI	$p = 0.043$
Octanal	12.60	1299	21.02 ± 0.01	25.83 ± 26.67	56.59 ± 23.91	91	MS/KI	$p = 0.083$
Nonanal	14.61	1393	80.98 ± 0.04	94.84 ± 1.20	97.52 ± 56.52	91	MS/KI	$p = 0.001$
Ketones								
3-Octanone	11.87	1266	23.44 ± 0.02	36.39 ± 46.46	66.98 ± 31.91	95	MS/KI	$p = 0.073$
Heterocyclic								
Benzothiazole	23.27	1954	17.24 ± 0.02	24.53 ± 0.66	45.52 ± 0.30	91	MS/KI	$p = 0.062$
Benzene derivatives								
Toluene	6.57	1027	381.95 ± 0.06	221.99 ± 51.56	735.19 ± 84.40	91	MS/KI	$p = 0.099$
o-Xylene	9.13	1164	25.02 ± 0.10	27.24 ± 30.36	72.86 ± 9.48	95	MS/KI	$p = 0.114$
p-Cymene	12.18	1280	29.35 ± 0.07	103.31 ± 15.33	181.68 ± 62.54	95	MS/KI	$p = 0.139$
Hydrocarbons								
2,2,4,6,6-pentamethyl-heptane	3.86	<800	171.27 ± 26.42	226.12 ± 360.83	435.97 ± 196.77	83	MS	$p = 0.074$
Sulfur compounds								
Carbon disulfide	1.62	<800	825.58 ± 359.41	125.31 ± 44.37	376.24 ± 44.34	90	MS	$p = 0.168$
Ethers								
15-Crown ether	30.23	2432	5.11 ± 0.01	5.77 ± 2.82	20.41 ± 14.99	90	MS/KI	$p = 0.158$

VOCs: volatile compounds. RT: retention time. SD: standard deviation. Every value is the average of two determinations ($n = 2$). MSLM: mass spectral library match. MI: method of identification, with reference to mass spectra (MS) provided by Wiley 275 MS database and Kovats indices (KI). *p*: significance as assessed by paired samples *t*-test, at the confidence level ≤ 0.05.

Considerable variations in the volatile profile of raw unprocessed meat have been reported previously in numerous studies, involving the volatile profile of the raw meat of duck, goose, pork [1], beef [14,17] and lamb [2,15,19]. What is worth mentioning is that the collective results of breed comparisons suggested that genetic effects on lamb flavour were minor as compared with the effects of other factors such as age, sex, feeding/diet, oxidation, lipid content, myoglobin and pH [20]. In these factors mentioned previously in the literature, the role of air packaging technology, in terms of real market analysis, in the development of raw lamb's meat aroma during storage should be also investigated. In real market analysis, consumers often purchase fresh, unprocessed meat from supermarkets or butcher's shops, packaged or covered with paper roll, and store it under refrigeration prior to cooking, roasting or grilling. In that sense, the development of aroma in packaged lamb meat during storage at 4 ± 1 °C has not been yet investigated. The specific trend in lamb meat aroma evolution, along with the characteristics/role of each class compound determined in the present study, follows the text sequence.

3.1. Alcohols

Garcia et al. [21] and Barbieri et al. [22] indicated that 1-pentanol and 1-hexanol, were produced from the degradation of homologous aldehydes during lipid and amino acid oxidation. 1-Hexanol has a herbal and fatty odour, whereas 1-pentanol has a pleasant, sweet or fruity odour [20]. Respective odour thresholds are 2500 and 10,000 ng/g [23].

1-Octen-3-ol has been reported to be a common product of oxidation. Several precursors have been reported for 1-octen-3-ol, including linoleic and arachidonic acids [24]. In a recent study, 1-octen-3-ol was found at higher amounts in the meat of lambs fed with olive cake as compared to those fed with linseed [2]. It is characterized by a mushroom-like, grassy odour [20]. Its odour threshold has been reported to be 1 ng/g [23].

In accordance with other researchers, the alcohols identified in the present study have been previously reported in studies involving fresh meat of pork, duck, goose [1], beef [14,17], turkey [11], cattle [13] and lamb [2,15], respectively. We may then conclude that these alcohols are typical volatile markers of raw and unprocessed meat.

3.1.1. Aldehydes

Aldehydes, in general, are the main representatives of volatile compounds derived from ruminant animal's meat [25]. According to Mottram [26], the straight chain aldehydes are compounds derived from the oxidation of fat. Hexanal, 2-heptenal, and 2,4-decadienal, (derived from linoleic acid), were associated with a diet rich in small amounts of plants [27,28]. Aldehydes act as important components in the transmission of certain flavors. The short-chain aldehydes tend to have a pungent or acidic odour, with the increasing of fatty acid chain length, and the degree of unsaturation.

Among aldehydes that are secondary products of fat oxidation [29], hexanal is a product resulting from the oxidation of unsaturated omega-6 fatty acids, having a low odour detection limit of 5.87 ng/g [30] and a paint-like, herbal, and rancid taste [31]. Heptanal has been reported to have a fatty odour, whereas octanal and nonanal a soapy or waxy odour [20,32]. The most abundant aldehydes identified in raw lamb meat were hexanal, followed by nonanal, heptanal, and octanal.

It should also be mentioned that the aforementioned aldehydes have been reported previously, adding to the volatile profile of lamb meat packaged under aerobic conditions [2], in agreement with present results.

3.1.2. Ketones

Ketones possess a low detection odour threshold [14]. The only ketone that was identified in raw lamb meat was 3-octanone. This volatile compound has been associated with the animal breed or the type of animal feeding background [13,14]. In addition, 3-octanone was found to be favorable to thermal processing of raw, trimmed of fat, cattle meat [13]. It is characterized by a fruity odour and an odour threshold ranging between 16–28 ng/g [33].

3.1.3. Hydrocarbons: Aliphatic and Benzene Derivatives

The hydrocarbons identified (aliphatic and aromatic) may originate from the animal food origin. These compounds are absorbed in fat tissue [34], as they are obtained from the degradation of lipids [21]. The aliphatic hydrocarbon 2,2,4,6,6-pentamethyl-heptane has been reported previously, aiding to the volatile profile of raw cattle meat [13,14] and lamb meat [2] having a characteristic odour [35].

On the other hand, there is a class of aromatic hydrocarbons that is characterized by a high degree of unsaturation and unusual stability. The most common member of this class is benzene (C_6H_6). When one of the positions on the ring has been substituted with another atom or group of atoms, the resulting compound is a mono-substituted benzene. The more common ones are toluene, *o*-xylene, and *p*-cymene. In the present study, these benzene derivatives contributed to the volatile pattern of raw lamb meat, recording a high rate of evolution, especially toluene and *p*-cymene, during storage (Table 1). This is also in agreement with the work of Insausti et al. [14]. However, their overall variation during storage was insignificant ($p > 0.05$) (Table 1). Some researchers have reported that aromatic hydrocarbons may be formed during the thermal decomposition of hydrocarbons, fats, and proteins [36]. However, in the present study these volatile compounds were found in raw, not thermally treated lamb meat. This is in agreement with the results of Insausti et al. [14], Saraiva et al. [17] and King et al. [37].

The presence/and or the combination of such mono-substituted benzenes, may play an important role in the overall flavour development of raw lamb meat, even though no individual compound of this group has a characteristic meat-like odour [38]. This is also in agreement with the results of Min et al. [39]. The typical odour of toluene and *o*-xylene is sweet, pungent or paint-like [38,40]. Toluene has an odour threshold of 2900 ng/g whereas that of *o*-xylene ranges between 620–1000 ng/g [40].

Insausti et al. [14] and Vasta et al. [13,15] documented the contribution of such compounds to the overall aroma development of raw beef, cattle and lamb meat, respectively.

3.1.4. Sulfur Compounds

Generally, sulfur compounds arise from sulfur-containing amino acids, cysteine and methionine, produced by proteolytic enzymes found in psychrotrophic bacteria (usually present in slaughterhouses). In the present study, carbon disulfide and benzothiazole were the only sulfur compounds determined. Carbon disulfide (CS_2) may also be derived from dithio-carbamate fungicides used in agriculture [41]. Carbon disulfide has a pleasant, sweet or ether-like odour [42] and an odour threshold ranging between 1.6–420 ng/g [23]. In the study of Saraiva et al. [17] carbon disulfide contributed to the overall aroma of raw beef packaged under vacuum or modified atmosphere packaging during refrigerated storage.

On the other hand, benzothiazole, is an aromatic heterocyclic compound found commonly in nature [43] and has an odour threshold of 80 ng/g [23]. It has been reported to have an unpleasant, dirty, metallic or pyridine-like odour [44]. The thiazole derivatives have been identified among the volatile components of coffee, boiled meat, boiled potatoes, sterilized milk and beer. Thiazoles and thiazolines are mainly formed by Strecker degradation of sulfur amino acids (particularly cysteine). Lorenz et al. [45] found that sulfur compounds were the main components of meat extract. Various studies indicated that sulfur compounds were present at higher amounts in animals which grazed normally, as compared to animals fed with concentrated food. In that sense, Young et al. [28] indicated that dimethyl-sulfide and dimethyl-sulfone were detected in lamb fat and were associated with pasture activities. The sulfur compounds in meat arise during cooking by the reaction of hydrogen sulfide (formed by cleavage of the amino acid cysteine) with carbonyl compounds derived from the Maillard reaction [26,46]. Given that lamb meat was not thermally treated, carbon disulfide and benzothiazole could be considered as typical volatiles of raw lamb meat associated with pasture diets/activities.

3.1.5. Ethers

Vapors of certain ethers may be used as insecticides, miticides, and fumigants for soil [47]. The only ether that was identified in the volatile fraction of raw lamb meat was 15-crown ether, recording insignificant variations ($p > 0.05$) with respect to storage time. Its presence in the volatile pattern of unprocessed lamb meat may be attributed to environmental contamination.

4. Discussion

The volatile fraction of raw lamb meat was largely dominated by aldehydes, which are secondary products of oxidation and have been considered as markers of fat oxidation during storage time [48]. However, several volatiles that are not directly related to lamb meat flavour were also identified, namely benzothiazole, *o*-xylene, toluene, and crown ether-15. Such compounds may be considered as artifacts originating from environmental or external contamination and may not be considered as "natural" or unambiguous meat volatiles.

The evolution of aldehydes with respect to storage time recorded positive Pearson's correlation (r) ($p < 0.05$). Respective r values for the groups hexanal–nonanal, heptanal–nonanal, and octanal–nonanal were 0.82, 0.44, and 0.72, through the 9 days of storage. However, the groups hexanal–nonanal and octanal–nonanal recorded excellent r values equal to 1 for the first 5 days of storage, in which raw lamb meat retained a fresh character based on microbiological (total viable counts < 7 log CFU/g) and sensory data (odour and taste) [18]. This was not the case for the group heptanal–nonanal, in which a perfect negative Pearson's correlation was recorded ($r = -1$). The evolution of the aforementioned aldehydes through the 9 days of storage is given in Table 2.

Table 2. Aldehyde ratios and malonic dialdehyde development with respect to storage time (days).

Aldehyde Ratio-MDA	Day 1	Day 5	Day 9	Pearson's (r)
Hexanal–Nonanal	2.02	2.44	3.49	1.00
Heptanal–Nonanal	0.45	0.31	0.62	0.87
Octanal–Nonanal	0.26	0.27	0.58	0.83
Sum of (Hexanal, Heptanal, Octanal)–Nonanal	2.73	3.02	4.70	0.89
MDA (mg/kg)	1.4	2.8	3.8	

Aldehyde ratios were obtained by dividing average values of each aldehyde presented in Table 1. MDA: malonic dialdehyde. Pearson's correlation coefficient r was considered at the confidence level $p < 0.05$.

Specific aldehyde ratios, namely hexanal to nonanal (Hex/Non), heptanal to nonanal (Hept/Non), octanal to nonanal (Oct/Non), and sum of hexanal, heptanal, and octanal to nonanal ratio (Hex + Hept + Oct/Non) (Table 2) recorded strong positive Pearson's correlation with the degree of oxidation values (TBA test) expressed as mg malonic dialdehyde per kg of lamb meat (mg MDA/kg) throughout storage (1.4 mg MDA/kg on day 1, 2.8 mg MDA/kg on day 5, and 3.8 MDA/kg on day 9) [18]. Respective Pearson's correlation values were $r = 1.00$, $r = 0.87$, $r = 0.83$, and $r = 0.89$ (Table 2). It should be stressed that the hexanal to nonanal ratio has been proposed as an indicator of the level of oxidation in olive oil [49], whereas variations in hexanal content with respect to storage time could serve as indicators of flavour deterioration in cooked pork [48]. In addition, octanal and nonanal (among other volatiles) have been proposed as indicators of raw beef quality stored under refrigeration [17].

Based on the aforementioned, the ratios Hex/Non, Oct/Non, and Hex + Hept + Oct/Non may be proposed as indicators of freshness and degree of oxidation, enhancing the quality of fresh lamb meat stored under refrigeration.

5. Conclusions

Results of the present study showed that the aroma of raw, unprocessed lamb meat is the outcome of the combination of specific volatile compounds. Storage time proved to be a parameter that significantly affected the evolution of hexanal, heptanal, and nonanal ($p < 0.05$). Furthermore, the evolution of hexanal, heptanal, octanal, and nonanal during storage under refrigeration, along with the Hex/Non, Oct/Non, and Hex + Hept + Oct/Non ratios could be proposed as overall indicators of lamb meat freshness and degree of oxidation, since they were strongly correlated with shelf life and TBA test data.

Special attention should be given to hexanal to nonanal ratio, which could serve as a unique marker of freshness and overall quality, since was perfectly correlated with the results of degree of oxidation (TBA test). Therefore, this ratio value—equal or lower than 2.45—is proposed as an indicator of lamb meat freshness and overall quality.

As an executive summary, the present study, unique in the literature, aids the fast quality control of raw lamb meat that enters the market by proposing specific aldehydes or aldehyde ratios as markers of raw lamb meat quality, in support of some previous studies on other food products [17,48,49].

Acknowledgments: The author would like to thank Michael G. Kontominas who provided access to the Laboratory of Food Chemistry, Department of Chemistry, University of Ioannina, and Anastasia Badeka for her technical assistance during preparation of samples for the GC/MS analysis. SVEKI SA is also acknowledged for the donation of lamb meat.

Conflicts of Interest: The author declares no conflict of interest.

Funding: This research did not receive any specific grant from funding agencies in the public, commercial, or not-for-profit sectors.

References

1. Soncin, S.; Chiesa, L.M.; Cantoni, C.; Biondi, P.A. Preliminary study of the volatile fraction in the raw meat of pork, duck and goose. *J. Food Comp. Anal.* **2007**, *20*, 436–439. [CrossRef]
2. Gravador, R.S.; Serra, A.; Luciano, G.; Pennisi, P.; Vasta, V.; Mele, M.; Pauselli, M.; Priolo, A. Volatiles in raw and cooked meat from lambs fed olive cake and linseed. *Animal* **2015**, *9*, 715–722. [CrossRef] [PubMed]
3. Estevez, M.; Morcuende, D.; Ventanas, S.; Cava, R. Analysis of volatiles in meat from Iberian pigs and lean pigs after refrigeration and cooking by using SPME-GC-MS. *J. Agric. Food Chem.* **2003**, *51*, 3429–3435. [CrossRef] [PubMed]
4. Acevedo, C.A.; Creixell, W.; Pavez-Barra, C.; Sánchez, E.; Albornoz, F.; Young, M.E. Modeling volatile organic compounds released by bovine fresh meat using an integration of solid phase microextraction and databases. *Food Bioprocess Technol.* **2012**, *5*, 2557–2567. [CrossRef]
5. Montel, M.C.; Masson, F.; Talon, R. Bacterial role in flavour development. *Meat Sci.* **1998**, *49*, S111–S123. [CrossRef]
6. Brunton, N.P.; Cronin, D.A.; Monahan, F.J. The effects of temperature and pressure on the performance of Carboxen/PDMS fibres during solid phase microextraction (SPME) of headsapace volatiles from cooked and raw turkey breast. *Flavour Fragr. J.* **2001**, *16*, 294–302. [CrossRef]
7. Moon, S.Y.; Cliff, M.A.; Li-Chan, E.C.Y. Odour-active components of simulated beef flavour analysed by solid phase microextraction and gas chromatography-mass spectrometry and-olfactometry. *Food Res. Int.* **2006**, *39*, 294–308. [CrossRef]
8. Pérez, R.A.; Rojo, M.D.; González, G.; De Lorenzo, C. Solid-phase microextraction for the determination of volatile compounds in the spoilage of raw ground beef. *J. AOAC Int.* **2008**, *91*, 1409–1415. [PubMed]
9. Verzera, A.; Dima, G.; Tripodi, G.; Ziino, M.; Lanza, C.M.; Mazzaglia, A. Fast quantitative determination of aroma volatile constituents in melon fruits by headspace-solid-phase microextraction and gas chromatography-mass spectrometry. *Food Anal. Methods* **2011**, *4*, 141–149. [CrossRef]
10. Condurso, C.; Tripodi, G.; Cincotta, F.; Lanza, C.M.; Mazzaglia, A.; Verzera, A. Quality assessment of Mediterranean shrimps during frozen storage. *Ital. J. Food Sci.* **2016**, *28*, 497–509.
11. Ahn, D.U.; Jo, C.; Olson, D.G. Volatile profiles of raw and cooked turkey thigh as affected by purge temperature and holding time before purge. *J. Food Sci.* **1999**, *64*, 230–233. [CrossRef]
12. Nam, K.C.; Ahn, D.U. Combination of aerobic and vacuum packaging to control lipid oxidation and off-odor volatiles of irradiated raw turkey breast. *Meat Sci.* **2003**, *63*, 389–395. [CrossRef]

13. Vasta, V.; Ratel, J.; Engel, E. Mass spectrometry analysis of volatile compounds in raw meat for the authentication of the feeding background of farm animals. *J. Agric. Food Chem.* **2007**, *55*, 4630–4639. [CrossRef] [PubMed]
14. Insausti, K.; Beriain, M.J.; Gorraiz, C.; Purroy, A. Volatile compounds of raw beef from 5 local Spanish cattle breeds stored under modified atmosphere. *J. Food Sci.* **2002**, *67*, 1580–1589. [CrossRef]
15. Vasta, V.; D'Alessandro, A.G.; Priolo, A.; Petrotos, K.; Martemucci, G. Volatile compound profile of ewe's milk and meat of their suckling lambs in relation to pasture vs. indoor feeding system. *Small Rumin. Res.* **2012**, *105*, 16–21. [CrossRef]
16. Gąsior, R.; Wojtycza, K. Sense of smell and volatile aroma compounds and their role in the evaluation of the quality of products of animal origin—A review. *Ann. Anim. Sci.* **2016**, *16*, 3–31. [CrossRef]
17. Saraiva, C.; Oliveira, I.; Silva, J.A.; Martins, C.; Ventanas, J.; García, C. Implementation of multivariate techniques for the selection of volatile compounds as indicators of sensory quality of raw beef. *J. Food Sci. Technol.* **2015**, *52*, 3887–3898. [CrossRef] [PubMed]
18. Karabagias, I.; Badeka, A.; Kontominas, M.G. Shelf life extension of lamb meat using thyme or oregano essential oils and modified atmosphere packaging. *Meat Sci.* **2011**, *88*, 109–116. [CrossRef] [PubMed]
19. Elmore, J.S.; Cooper, S.L.; Enser, M.; Mottram, D.S.; Sinclair, L.A.; Wilkinson, R.G.; Wood, J.D. Dietary manipulation of fatty acid composition in lamb meat and its effect on the volatile aroma compounds of grilled lamb. *Meat Sci.* **2005**, *69*, 233–242. [CrossRef] [PubMed]
20. Calkins, C.R.; Hodgen, J.M. A fresh look at meat flavor. *Meat Sci.* **2007**, *77*, 63–80. [CrossRef] [PubMed]
21. Garcia, C.; Berdague, J.L.; Antequera, T.; López-Bote, C. Volatile components of dry-cured Iberian ham. *Food Chem.* **1991**, *41*, 23–32. [CrossRef]
22. Barbieri, G.; Bolzoni, L.; Parolari, G.; Virgili, R.; Buttini, R.; Careri, M.; Mangia, A. Flavour compounds of dry-cured ham. *J. Agric. Food Chem.* **1992**, *40*, 2389–2394. [CrossRef]
23. Leffingwell and Associates. 2017. Available online: http://www.leffingwell.com/ (accessed on 10 October 2017).
24. Wilson, R.A.; Katz, I. Review of literature on chicken flavour and report of isolation of several new chicken flavour components from aqueous cooked chicken broth. *J. Agric. Food Chem.* **1972**, *20*, 741–747. [CrossRef]
25. Larick, D.K.; Turner, B.E. Headspace volatiles and sensory characteristics of ground beef from forage- and grain-fed heifers. *J. Food Sci.* **1990**, *54*, 649–654. [CrossRef]
26. Mottram, D.S. Flavour formation in meat and meat products: A review. *Food Chem.* **1998**, *62*, 415–424. [CrossRef]
27. Raes, K.; Balcaen, A.; Dirinck, P.; De Winne, A.; Claeys, E.; Demeyer, D.; De Smet, S. Meat quality, fatty acid composition and flavour analysis in Belgian retail beef. *Meat Sci.* **2003**, *65*, 1237–1246. [CrossRef]
28. Young, O.A.; Lane, G.A.; Priolo, A.; Fraser, K. Pastoral and species flavour in lambs raised on pasture, lucerne or maize. *J. Sci. Food Agric.* **2003**, *83*, 93–104. [CrossRef]
29. Van Ruth, S.M.; Cheraghi, T.; Roozen, J.P. Lipid-derived off flavours in meat by-products: Effect of antioxidants and Maillard reactants. In *Flavour of Meat, Meat Products and Seafood*; Shahidi, F., Ed.; Blackie Academic & Professional: London, UK, 1998; p. 260. ISBN 978-0-7514-0484-5.
30. Ajuyah, A.O.; Fenton, T.W.; Hardin, R.T.; Sim, J.S. Measuring lipid oxidation volatiles in meats. *J. Food Sci.* **1993**, *58*, 270–273. [CrossRef]
31. Brewer, M.S.; Vega, J.D. Detectable odor thresholds of selected lipid oxidation compounds in a meat model system. *J. Food Sci.* **1995**, *60*, 592–595. [CrossRef]
32. Acree, T.; Aru, H. Flavornet. 1997. Available online: http://www.nysaes.cornell.edu/flavornet/chem.html (accessed on 30 October 2001).
33. Wasowicz, E.; Kaminski, E.; Kollmannsberger, H.; Nitz, S.; Berger, G.; Drawert, F. Volatile components of sound and musty wheat grains. *Chem. Mikrobiol. Technol. Lebensm.* **1988**, *11*, 161–168.
34. Meynerier, A.; Novelli, E.; Chizzolini, R.; Zanardi, E.; Gandemer, G. Volatile compounds of commercial Milano salami. *Meat Sci.* **1999**, *51*, 175–183. [CrossRef]
35. Gorraiz, C.; Beriain, M.J.; Insausti, K. Effect of aging time on volatile compounds, odor, and flavor of cooked beef from Pirenaica and Friesian Bulls and Heifers. *J. Food Sci.* **2002**, *67*, 916–922. [CrossRef]
36. Cha, Y.J.; Back, H.H.; Hsieh, T.C.Y. Volatile components in flavor concentrates from crayfish processing waste. *J. Sci. Food Agric.* **1992**, *59*, 239–248. [CrossRef]

37. King, M.F.; Hamilton, B.L.; Mattews, M.A.; Rule, D.C.; Field, R.A. Isolation and identification of volatiles and condensable material in raw beef with supercritical carbon dioxide extraction. *J. Agric. Food Chem.* **1993**, *41*, 1974–1981. [CrossRef]
38. Molo, L.; Dekimpe, J.; Etlevant, P.; Addeo, F. Neutral volatile compounds in the raw milks from different species. *J. Dairy Res.* **1993**, *60*, 199–213.
39. Min, D.B.S.; Ina, K.; Peterson, R.J.; Chang, S.S. Preliminary identification of volatile flavor compounds in the neutral fraction of roasted beef. *J. Food Sci.* **1979**, *44*, 639–642. [CrossRef]
40. The National Institute for Occupational Safety and Health. 2007. Available online: http://www.cdc.gov/niosh (accessed on 15 September 2007).
41. Kontou, S.; Tsipi, D.; Tzia, C. Stability of the dithiocarbamate pesticide maneb in tomato homogenates during cold storage and thermal processing. *Food Addit. Contam.* **2004**, *21*, 1083–1089. [CrossRef] [PubMed]
42. Holleman, A.F.; Wiberg, E.; Wiberg, N. *Inorganic Chemistry*; Academic Press: Cambridge, MA, USA, 2001; ISBN 100123526515.
43. Le Bozec, L.; Moody, C.J. Naturally occurring nitrogen–sulfur compounds. The benzothiazole alkaloids. *Aust. J. Chem.* **2009**, *62*, 639–647. [CrossRef]
44. Lewis, R.J. *Hawley's Condensed Chemical Dictionary*, 12th ed.; Van Nostrand Reinhold, Co.: New York, NY, USA, 1993; p. 132.
45. Lorenz, G.; Stern, D.J.; Flath, R.A.; Haddon, W.F.; Tillin, S.J.; Teranishi, R. Identification of sheep liver volatiles. *J. Agric. Food Chem.* **1983**, *31*, 1052–1057. [CrossRef]
46. Hoffmann, T.; Schieberle, P.; Grosch, W. Model studies on the oxidative stability of odor-active thiols occurring in food flavors. *J. Agric. Food Chem.* **1996**, *44*, 251–255. [CrossRef]
47. Wade, L.G. *Ether*; Encyclopædia Britannica, Inc.: Chicago, IL, USA, 2017.
48. Shahidi, F.; Pegg, R.C. Hexanal as an indicator of meat flavor deterioration. *J. Food Lipids* **1994**, *1*, 177–186. [CrossRef]
49. Robertson, G.L. *Food Packaging and Shelf Life: A Practical Guide*; CRC Press, Taylor and Francis Group: Boca Raton, FL, USA, 2009; p. 328. ISBN 9781420078442.

Article

A Comparison Study of Quality Attributes of Ground Beef and Veal Patties and Thermal Inactivation of *Escherichia coli* O157:H7 after Double Pan-Broiling Under Dynamic Conditions

KaWang Li [1], Amanda Gipe McKeith [2], Cangliang Shen [1,*] and Russell McKeith [3]

[1] Division of Animal and Nutritional Sciences, West Virginia University, Morgantown, WV 26506, USA; kwli@mix.wvu.edu

[2] Department of Animal Sciences & Agricultural Education, California State University Fresno, Fresno, CA 93740, USA; amckeith@mail.fresnostate.edu

[3] Division of Agriculture, College of the Sequoias, Tulare, CA 93274, USA; russellm@cos.edu

* Correspondence: cashen@mail.wvu.edu; Tel.: +1-304-293-2691

Received: 16 October 2017; Accepted: 22 December 2017; Published: 26 December 2017

Abstract: This study compared the quality variation and thermal inactivation of *Escherichia coli* O157:H7 in non-intact beef and veal. Coarse ground beef and veal patties (2.1 cm thick, 12.4 cm diameter, 180 g) inoculated with *E. coli* O157:H7, aerobically stored before double pan-broiling for 0–360 s without rest or to 55, 62.5, 71.1, and 76 °C (internal temperature) with 0.5- or 3.5-min rest. Microbial population and qualities including color, cooking losses, pH, water activity, fat, and moisture content, were tested. After cooking the beef and veal patties, the weight losses were 17.83–29%, the pH increased from 5.53–5.60 to 5.74–6.09, the moisture content decreased from 70.53–76.02% to 62.60–67.07%, and the fat content increased ($p < 0.05$) from 2.19–6.46% to 2.92–9.45%. Cooking beef and veal samples with increasing internal temperatures decreased a* and b* values and increased the L* value. *Escherichia coli* O157:H7 was more sensitive to heat in veal compared to beef with shorter D-value and "shoulder" time. Cooking to 71.1 and 76 °C reduced *E. coli* O157:H7 by >6 log CFU/g regardless of rest time. Cooking to 55 °C and 62.5 °C with a 3.5-min rest achieved an additional 1–3 log CFU/g reduction compared to the 0.5-min rest. Results should be useful for developing risk assessment of non-intact beef and veal products.

Keywords: *Escherichia coli* O157:H7; quality; beef; veal; thermal inactivation

---

1. Introduction

Escherichia coli O157:H7 can generate shiga toxins that can cause, with as few as 10 cells, severe hemolytic uremic syndrome in infected humans [1]. *E. coli* O157:H7 has been considered an adulterant of raw, non-intact beef products since 1999 [2]. The United States Department of Agriculture, Food Safety and Inspection Service (USDA-FSIS) defines non-intact beef products as products that have gone through treatments, such as grinding, restructuring, or mechanical tenderization processes, including cubing, needling, and pounding [3]. These non-intact beef products have been involved in several *E. coli* O157:H7 outbreaks in the United States since 2000 [4]. During non-intact beef production, pathogen cells, such as those of *E. coli* O157:H7, on the meat surface may be translocated and trapped in sterile internal tissues, thus protected from thermal destruction if the meat is undercooked. A recent survey showed that 40–58% of US consumers ordered beefsteaks at medium rare (60–62.8 °C) to rare (54.4–57.2 °C), which could potentially put consumers at a high risk from *E. coli* O157:H7 contaminated non-intact veal meat if consumers order the same way as beefsteaks [5].

Thermal processing includes using high temperature to inactive spoilage and foodborne pathogens is one of the most effective and widely used technologies for meat products preservation [6]. The effectiveness of cooking in inactivating *E. coli* O157:H7 contaminated non-intact beef has been documented in numerous studies [4,7,8] indicating that the cooking effectiveness on pathogen inactivation increased in the order of broiling > grilling > frying, the thicker the products the higher reduction achieved, and lower fat content increased thermal inactivation activity.

Veal, which originated from Europe, is the meat from 16–18-week-old calves. In the past 10 years, 25% of American households purchased veal products in restaurants or retail stores at least once every three months [9]. Different veal cuts, such as cutlet, loin, rib, breast, and shank, are more popular to restaurant consumers due to their unique tenderness and flavor. Moreover, the nutrition of veal products matches the dietary guidelines that are recommended by the American Heart Association, the American Dietetic Association, and the USDA. The veal market generates approximately $1.5 billion sales each year in the US [10]. Although veal products have not been implicated in *E. coli* outbreaks in the US, since 2009 there have been multiple recalls of veal products amounting to 14,600 lb (ca. 6649 kg) due to possible *E. coli* O157:H7 and STEC contamination [11]. According to the USDA-FSIS, in May 2017, a large veal processor recalled over 5000 lb of ground veal, pork, and beef due to possible non O157:H7 shiga toxin producing *E. coli* contamination [12]. According to the USDA-FSIS, there is a greater prevalence of STEC in veal products than in other beef products. For example, in 2013, the USDA-FSIS in their testing of raw ground beef component samples in federal meat-processing factories discovered 0 (0%) of 733 samples to be positive for *E. coli* O157:H7 and three (0.24%) of 1232 samples to be positive for STEC in beef; in contrast, in veal, three (3.49%) of 86 samples were positive for *E. coli* O157:H7 and 4 (4.00%) of 100 samples were positive for STEC [13]. The difference in confirmed STEC-positive samples of veal compared to those of beef is striking and raises the question of whether the consumption of veal poses a greater risk to public health than that of beef. Currently, only two studies have reported the thermal inactivation of *E. coli* O157:H7 strains in non-intact veal products [11,14].

The safety of beef and veal products is important to the industry and to consumers, but consumers tend to identify the quality of products based on appearance. Cornforth and Jayasingh [15] stated that color is one of the most important characteristics regarding consumers' purchasing decisions, even though color is sometimes poorly related to meat quality. Fresh beef or veal meat is often displayed in styrofoam trays and covered with poly-vinyl chloride (PVC) oxygen-permeable films, which allow the rapid development of the desirable bright cherry-red (beef) or light pink color (veal), respectively, due to rapid pigment oxygenation. However, discoloration often occurs within 1 week of shelf time. Currently, the number of studies that focus on the quality changes in veal products during processing, storage and cooking in terms of factors such as water activity, pH, moisture, fat content and color change is very limited.

The objective of this study is to investigate the quality variances, including color variation in non-intact coarse ground beef and veal patties during aerobic storage and cooking and to evaluate the thermal inactivation of *E. coli* O157:H7 in coarse ground beef and veal patties. We hypothesize that (1) beef and veal patties have similar tendencies in quality change throughout storage and cooking and (2) a higher internal temperature with a longer rest time will increase the inactivation of *E. coli* O157:H7 in beef and veal patties. The novelty of this study are (1) a detailed side-by-side comparison study of quality attributes and thermal inactivation activity of *E. coli* O157:H7 between beef and veal and (2) the thermal kinetics study was conducted in a commercial size patties cooked on a griller instead of using small amount of meat heated in water bath.

2. Materials and Methods

2.1. Preparation of Bacterial Strains and Inoculum

Escherichia coli O157:H7 strains ATCC 43895, ATCC 43888, and ATCC 43889 (kindly provided by Beth Whittam, Michigan State University, East Lansing, MI, USA) were cultured and sub-cultured individually in 10 mL of tryptic soy broth (TSB) at 35 °C for 24 h. The three cultures were then mixed and centrifuged (Eppendorf model 5810R, Brinkmann Instruments Inc., Westbury, NY, USA) at 4629× *g* for 15 min at 4 °C. The harvested cells were washed twice with 10 mL of phosphate-buffered saline (PBS), centrifuged as described above, and re-suspended in 30 mL of fresh PBS. The washed pathogen cells were 10-fold diluted in PBS to obtain an initial inoculum level of 8 log CFU/mL, and then 40 mL of this prepared inoculum was added into 2 kg of coarse ground beef or veal to reach the inoculation level of ~6 log CFU/g.

2.2. Preparation of Non-Intact Ground Veal and Beef Patties

Fresh beef knuckles and veal round top were purchased from a local meat retailer for each replicate. The meat was manually cut into trimmings and then coarse ground in a meat grinder (Gander Mountain #5 Electric Meat Grinder, Saint Paul, MN, USA) with a kidney plate (0.95 cm diameter). The ground meat was then mixed with 40 mL of the aforementioned *E. coli* O157:H7 inoculum cocktail in a bowl-lift stand mixer (Kitchen Aid Professional 600, Benton Harbor, MI, USA) at medium speed for two minutes to ensure an even distribution of the inoculum into the sample, which simulates *E. coli* O157:H7 contamination during the preparation of non-intact beef or veal products. A manual hamburger patty maker (Mainstays 6-ounce-patty maker, Walmart, Bentonville, AR) was then used to make beef or veal patties with 180 g of grounded meat. The beef/veal patties (2.1 cm thick and 12.4 cm diameter) were packaged aerobically in foam trays (20 × 25 cm, Pactiv, Lake Forest, IL, USA) with the absorbent pads, covered using air-permeable plastic film (Omni-film, Pliant Corporation, OH, USA) and stored at 4.0 °C for four days.

2.3. Cooking Beef or Veal Patty Samples

After four days of storage, the beef or veal patties were removed from their packages, weighed, and double pan-broiled in a Farberware grill (Farberware 4-in-1 Grill, Fairfield, CA, USA) with a set-up temperature of 177 °C (or 350 °F) (1) for 0, 30, 60, 90, 120, 180, 240, and 360 s with 0-min rest to determine the thermal dynamic parameters (i.e., D-value, "shoulder", α) (2) to an internal geometric target temperature of 55, 62.5, 71.1, or 76 °C, followed by a 0.5- or 3.5-min rest. The cooked patties were allowed to rest on the tray after cooking without any cover. Double broiling, also known as contact grilling, is when the food (usually meat, especially burger patties, chicken, and steaks) is cooked on both sides simultaneously by applying two cooking surfaces, from both the bottom and the top, greatly reducing the cooking time. A type-K thermocouple was attached to the geometric center of the patty to monitor the internal temperature throughout cooking using PicoLog (Pico Technology Ltd., Cambridge, UK), a real-time data-recording software [4,7,8]. The cooked meat rest on the tray were also monitored the internal temperature using the same type-K thermocouple. The meat quality test including cooking losses, color, pH, water activity, moisture, and fat content were conducted in a separate study using the uninoculated beef and veal samples with the same storage and cooking treatments and cooled to room temperature after cooking.

2.4. Color Measurement

The objective color of non-intact beef or veal patties was measured on each day of storage and after cooking to 55, 62.5, 71.1, or 76 °C (internal and external parts) using a portable spectrophotometer (HunterLab MiniScan EZ, Reston, VA, USA), with full spectral data being obtained as L* (lightness), a* (redness), and b* (yellowness), along with reflectance data [16]. For the external surface color measurement, an average value for L*, a*, and b* was determined from the mean of three random readings on the surface from three pieces of each treatment that was used for the color analysis. To measure the internal color of the cooked samples, the beef or veal patties were split transversely across the longitudinal axis to expose the center portion with three random readings from three pieces of each treatment.

2.5. Physical, Chemical and Microbiological Analyses

Cooking losses were determined by measuring the difference in patty weight before cooking and after cooking when the samples had cooled to room temperature. The pH of the meat homogenate was measured after microbial analysis using a digital pH meter (Fisher Scientific, Fair Lawn, NY, USA). The water activity (aw) indicates the availability of water for bacterial growth. The water activity of the uncooked and cooked samples was measured using an AquaLab water activity meter (model series 3, Decagon Devices Inc., Pullman, WA, USA). All of the samples were tested for fat and moisture content at the Meat Science Lab of the University of Illinois at Urbana– Champaign. For microbiological analysis, the individual uncooked or cooked beef or veal samples were transferred to a Whirl-Pak filter bag (1627 mL, 19 × 30 cm, Nasco, Modesto, CA, USA) with a 1:1 ratio of nutrient broth by weight and homogenized (Masticator, IUL Instruments, Barcelona, Spain) for 2 min. Serial 10-fold dilutions of each sample in PBS were surface-plated onto tryptic soy agar (Acumedia, Lansing, MI, USA) supplemented with 0.1% sodium pyruvate (Fisher Scientific, Fair Lawn, NY; TSAP) and MacConkey agar (Acumedia, Lansing, MI, USA) for the enumeration of total bacterial populations and *E. coli* O157:H7, respectively. Colonies were counted manually after incubation at 35 °C for 48 h. The samples below the detect limit of spread-plating were enriched at 35 °C for 48 h and streak-plated onto MacConkey agar to enrich any cells that were not recovered.

2.6. Data Analysis

The experiment was repeated twice, with three samples in each replicate in quality and microbial thermal inactivation studies. The quality parameters of beef and veal samples, including cooling losses, pH, water activity, fat and moisture content, were analyzed with a one-way ANOVA of SAS. All of the comparisons were performed with $p = 0.05$. Microbial populations (log CFU/g) were analyzed using the PROC MIXED procedure of Statistical Analysis System (SAS; version 9.3, SAS Institute Inc., Cary, NC, USA), with independent variables including beef or veal, cooked internal temperatures, rest time, and interactions between two or three variables. USDA-Integrated-Predictive-Modeling-Program software [17], provided by Dr. Lihan Huang, was used to estimate parameters of the survival of the pathogen cells in ground beef and veal samples during thermal processing with various heating time. The means and standard deviations were calculated, and the mean differences between treatments were determined using the Least Significant Difference (LSD) function for multiple comparisons at a significance level of $\alpha = 0.05$.

3. Results and Discussion

3.1. Cooking Curve and Weight Losses

The initial geometric center temperature of uncooked beef and veal patties ranged from 3.6 °C to 4.8 °C and from 4.1 °C to 8.9 °C, respectively. The cooking of beef samples by double pan-broiling required 330, 360, 430 and 460 s to reach the internal center temperatures of 55, 62.5, 71.1 and 76 °C, respectively (Figure 1A). In veal samples, it took 300, 330, 360, and 420 s to achieve internal temperatures of 55, 62.5, 71.1, and 76 °C, respectively (Figure 1B). The shorter cooking time that was required by the veal samples to reach the same internal temperatures compared to the beef samples is possibly to be explained by the following three reasons. First, the fiber density could be greater in veal than beef since veal is less mature than beef with the relatively lower muscle fiber content. Second, collagen immaturity and less fiber hypertrophy in the veal patties as compared to the more mature collagen and muscle fibers in beef muscle tissue allowing heat to transfer and penetrate the veal patties more efficiently. Third, the higher moisture content of veal could be a contributing factor to heating rate. As expected, during the 3.5-min resting time, in both beef and veal samples, the geometric center temperatures continued to increase from 61 °C to 65.9 °C, from 68.4 °C to 71.6 °C, and from 72 °C to 78.2 °C when cooking samples to 55, 62.5, and 71.1 °C, respectively (data not shown in tabular form). When cooking beef and veal samples to 76 °C, the temperature ranged from 74.6 °C to 78.5 °C and from 72.6 °C to 78 °C, respectively (data not shown in tabular form).

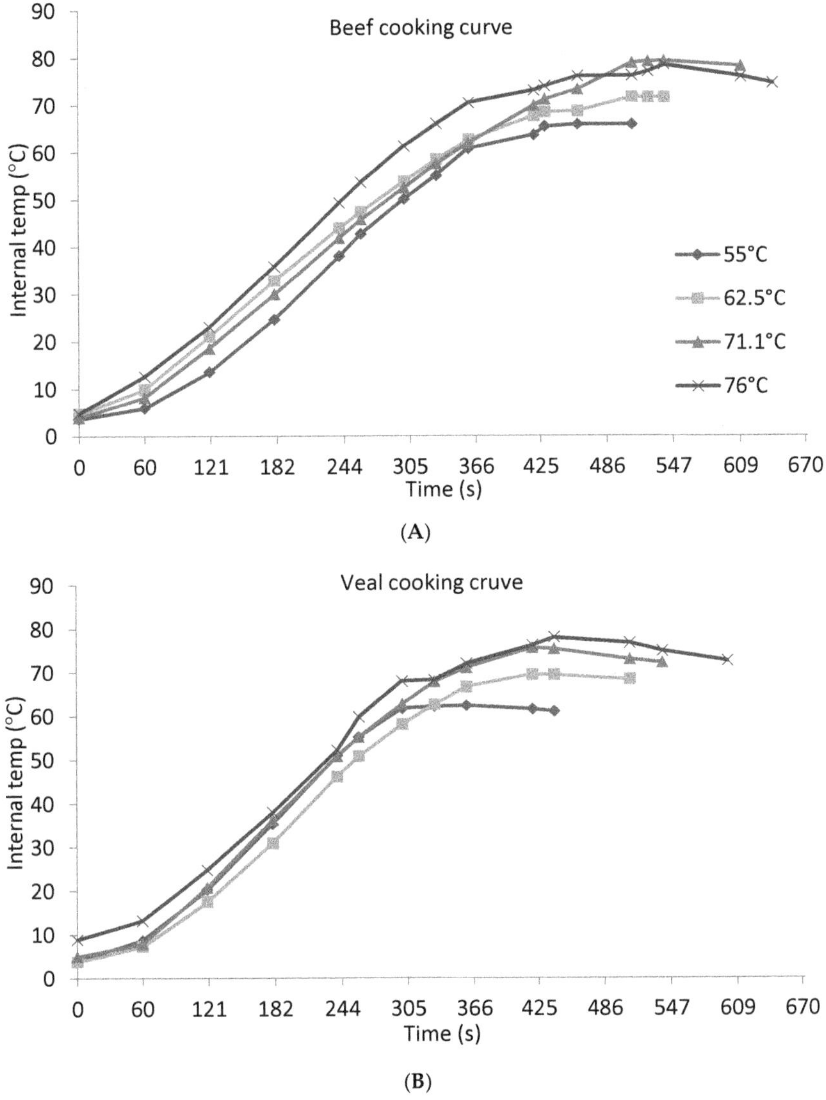

Figure 1. Cooking times and temperature curves for non-intact course ground beef (**A**) and veal (**B**) patties that were cooked by double pan-broiling using a Farberware grill.

3.2. Physical and Chemical Proprieties of Beef and Veal Samples

Cooking caused weight losses ranging from 17.83% to 29% in non-intact beef samples (Table 1) and from 19% to 29% in non-intact veal samples (Table 2). In beef samples, cooking to internal temperatures of 62.5 °C to 76 °C resulted in higher ($p < 0.05$) cooking losses (24.16–29%) compared to those from cooking to 55 °C (17.83%) (Table 1). In veal samples, double pan-broiling to internal temperatures of 71.1 °C and 76 °C resulted in higher ($p < 0.05$) cooking losses (28.25–29%) (Table 2). Higher cooked internal temperature resulted in higher cooking losses due to the prolonged cooking time, causing extra moisture loss via evaporation and the release of excess juice inside the meat samples.

The pH of uncooked beef and veal patties was 5.60 (Table 1) and 5.53 (Table 2), respectively. Double pan-broiling caused a significant increase ($p < 0.05$) in the pH of beef and veal patties, resulting in pH values ranging from 5.98 to 6.09 (Table 1) and from 5.73 to 5.78 (Table 2), respectively, in agreement with the previous studies [18,19]. The increase in pH for cooked meat is due to the reduction of free acidic groups as the meat temperature increases during heating [20]. However, no significant differences in the pH of cooked samples were observed when beef or veal patties were cooked to various internal target temperatures (55 °C to 76 °C). Only a slight pH increase from 5.98 to 6.09 was detected in beef samples after cooking from 55 °C to 76 °C.

While the moisture content describes the ratio of water mass to sample mass, water activity is the partial vapor pressure of pure water, which indicates the availability of water for bacterial growth. The water activity of fresh beef and veal patties was 0.992 (Table 1) and 0.991 (Table 2), respectively. In both beef and veal samples, the water activity did not change significantly after cooking to various internal temperatures. A previous study [21], which reported that cooking non-intact ground beef to internal temperatures of 60 °C and 65 °C resulted in water activities of 0.981 to 0.982 compared to the uncooked samples' value of 0.982 to 0.984 had similar results to this study. The initial moisture of beef and veal samples was 70.53% and 76.02%, respectively. In both beef and veal samples, the moisture content significantly decreased ($p < 0.05$) as the cooked internal temperature increased from 55 °C to 76 °C (Tables 1 and 2). Cooking beef or veal patties to 71.1 °C or 76 °C significantly decreased ($p < 0.05$) the moisture content to approximately 63% (beef) and 67–68% (veal) compared to the 66% (beef) and 71% (veal) in samples that were cooked to 55 °C (Tables 1 and 2). Previous studies [4,18] reported that the moisture content of ground beef patties and of moisture-enhanced reconstructed beef patties was lower after cooking. The decreased moisture content of beef and veal is likely due to a loss of water during cooking/heating [4].

The fat content of fresh beef and veal patties was 6.46% and 2.19% (Tables 1 and 2), respectively. Cooked beef samples had a significantly ($p < 0.05$) increased fat content of 8.62% to 9.45%, irrespective of the cooked internal temperatures (Table 1). Previous studies [4,19,21] reported that cooking low-fat ground beef or non-intact beef increased the fat content due to the moisture loss. A slightly ($p = 0.074$, >0.05) increased fat content ranging from 2.79% to 3.02% was found in cooked veal samples compared to that in the uncooked samples.

Table 1. Cooking losses, pH, water activity, and moisture and fat contents of non-intact course ground beef patties before and after cooking to various internal end-point temperatures.

Beef	Before Cooking	After Heating to (°C)			
		55	62.5	71.1	76
Cooking losses (%)	-	17.83 ± 5.56 ^{a}	24.17 ± 2.71 ^{b}	26.67 ± 2.50 ^{b}	29.00 ± 1.55 ^{b}
pH	5.60 ± 0.07 ^{a}	5.98 ± 0.11 ^{b}	6.07 ± 0.16 ^{b}	6.08 ± 0.15 ^{b}	6.09 ± 0.15 ^{b}
Water activity	0.992 ± 0.001 ^{a}	0.990 ± 0.003 ^{a}	0.991 ± 0.005 ^{a}	0.990 ± 0.003 ^{a}	0.987 ± 0.003 ^{a}
Moisture (%)	70.53 ± 0.55 ^{a}	66.63 ± 1.85 ^{b}	64.36 ± 1.23 ^{bc}	63.04 ± 1.25 ^{c}	62.60 ± 1.18 ^{c}
Fat (%)	6.46 ± 0.77 ^{a}	8.62 ± 0.65 ^{b}	9.34 ± 0.44 ^{b}	9.45 ± 0.55 ^{b}	8.96 ± 0.60 ^{b}

Mean values with a different lowercase letter within a row are significantly different ($p < 0.05$).

Table 2. Cooking losses, pH, water activity, and moisture and fat contents of non-intact course ground veal patties before and after cooking to various internal end-point temperatures.

Veal	Before Cooking	After Heating to (°C)			
		55	62.5	71.1	76
Cooking losses (%)	-	19.00 ± 3.56 a	20.75 ± 5.50 a	28.25 ± 0.96 b	29.00 ± 0.82 b
pH	5.53 ± 0.01 a	5.78 ± 0.02 b	5.74 ± 0.08 b	5.73 ± 0.08 b	5.74 ± 0.08 b
Aw	0.991 ± 0.005 a	0.989 ± 0.003 a	0.988 ± 0.002 a	0.987 ± 0.002 a	0.988 ± 0.001 a
Moisture (%)	76.02 ± 0.36 a	71.19 ± 0.51 b	69.57 ± 2.01 bc	67.95 ± 0.49 cd	67.07 ± 0.89 d
Fat (%)	2.19 ± 0.25 a	2.79 ± 0.54 a	3.00 ± 0.44 a	3.02 ± 0.46 a	2.92 ± 0.17 a

Mean values with a different letter within a row are significantly different ($p < 0.05$), Aw: water activity.

3.3. Color Variation during Storage and Cooking

The color index a*, b*, and L* values of freshly prepared beef patties were 34.75, 25.89, and 44.94, respectively (Table 3). Compared to beef samples, lower ($p < 0.05$) a* and b* values of 26.98 and 22.63 and a higher L* value of 59.83 were detected in fresh veal patties (Table 3). The less-red and lighter color is expected in veal samples because veal is the meat of bovine animals aged eight months or less, containing less myoglobin compared to the beef. During the aerobic storage, in general, the a*, b*, and L* values decreased ($p < 0.05$) from 34.75 to 15.27, from 25.86 to 13.93, and from 46.32 to 39.85 in beef patties and decreased ($p < 0.05$) from 26.98 to 12.77, from 22.63 to 16.53, and from 59.82 to 57.87 in veal samples by the end of storage. These results agree with those of a previous study [20], which found that the a*, b*, and L* values decreased in beef samples as the display time increased from 0 to three days. Madhavi and Carpenter [22] also reported that discoloration occurs within seven days of wrapping beef in oxygen-permeable film. During PVC film storage, oxymyoglobin reacted with oxygen to form metmyoglobin, causing the less-red color of the beef and veal samples.

After cooking to 55–76 °C, the external color values of a* and b* ranged from 11.86 to 13.46 and from 16.96 to 18.45 in beef samples (Table 4(A)) and from 10.03 to 11.15 and from 18.71 to 21.53 in veal samples (Table 4(A)). In both beef and veal samples, increasing the cooked internal temperature from 55 to 76 °C did not significantly change the values of a* and b*. During double pan-broiling, the external surfaces of the beef and veal samples were in close contacted with the heat surface of the grill, which cause the oxymyoglobin to quickly become metmyoglobin, producing a brownish color regardless of the cooked internal temperature. However, it is interesting to note that the L* value of the external surface was different between beef and veal samples. Cooking beef samples from 62.5 to 76 °C decreased ($p < 0.05$) the L* value from 46.67 to 47.08 compared to the value that was obtained at 55 °C (50.32). In veal samples, the L* value decreased ($p < 0.05$) from 68.93 to 66.45 when the cooked temperature increased from 55 to 71.1 °C, while the L* value returned to 68.45 after cooking to 76 °C. There results suggest that the doneness of cooked beef and veal patties cannot be determined by external color change.

In general, the a* and b* values of the internal color of cooked beef samples decreased (less red and yellow) (Table 4(B)) and the L* value increased as the internal end-point temperature increased (Table 4(B)). For the a* and b* values, lower values of 13.95 (a*, less red) and 17.93 (b*, less yellow) were detected in the beef samples that were cooked to 76 °C compared to the 28.05 (a*) and 24.41 (b*) of the samples that were cooked to 55 °C (Table 4(B)). However, the beef samples that were cooked to 62.5 °C or 71.1 °C had similar a* values of 19.34 to 21.75, and cooking to 55 °C or 62.5 °C resulted in similar b* values of 23.51 to 24.41 (Table 4(B)). For the L* value, cooking beef samples to 62.5, 71.1 and 76 °C resulted in a higher ($p < 0.05$) value of 53.66 to 54.19 compared to the value of 50.24 in samples that were cooked to 55 °C (Table 4(B)). Hague and others [23] reported that increasing the end-point cooking temperature from 55 °C to 77 °C decreased the a* and b* values of ground beef patties from 14.6 to 11.0 and from 18.4 to 15.9, respectively, and increased the L* value from 50.9 to 52.2. The variances in the internal cooked color were attributed to the denaturation of myoglobin in ground beef patties as the internal end-point temperature increased from 55 °C to 76 °C [24].

Table 3. Color values (L*, a*, and b*) of the external parts of course ground beef and veal patties stored aerobically at 4.0 °C for four days in foam trays that were covered with air-permeable plastic film.

	Beef					Veal				
	Day 0	Day 1	Day 2	Day 3	Day 4	Day 0	Day 1	Day 2	Day 3	Day 4
L*	46.32 ± 2.11 ^{a}	42.06 ± 2.75 ^{b}	40.20 ± 2.37 ^{c}	43.70 ± 2.71 ^{b}	39.85 ± 2.05 ^{c}	59.96 ± 1.64 ^{a}	59.16 ± 1.55 ^{a}	58.15 ± 2.44 ^{a}	58.26 ± 1.65 ^{a}	57.87 ± 1.64 ^{b}
a*	34.90 ± 2.27 ^{a}	22.92 ± 2.61 ^{b}	19.71 ± 2.26 ^{c}	15.40 ± 1.61 ^{d}	15.27 ± 1.02 ^{d}	26.90 ± 0.94 ^{a}	22.41 ± 1.38 ^{b}	16.97 ± 1.49 ^{c}	14.39 ± 1.04 ^{c}	12.77 ± 0.60 ^{d}
b*	25.86 ± 2.07 ^{a}	17.91 ± 1.62 ^{b}	17.14 ± 1.48 ^{b}	14.40 ± 1.60 ^{c}	13.93 ± 1.49 ^{c}	22.61 ± 0.95 ^{a}	20.07 ± 0.96 ^{b}	17.99 ± 0.95 ^{c}	17.30 ± 0.96 ^{c}	16.53 ± 0.83 ^{c}

Mean values with a different lowercase letter within a row within beef or veal are significantly different ($p < 0.05$).

Table 4. Color values (L*, a*, and b*) of the external (A) and internal (B) parts of course ground beef and veal patties double pan-broiling to internal end-point temperatures of 55, 62.5, 71.1, and 76 °C.

	L*		a*		b*	
Cooked Internal Temperature (°C)	Beef	Veal	Beef	Veal	Beef	Veal
(A) External parts						
55	50.32 ± 3.18 ^{aA}	68.98 ± 2.13 ^{aB}	13.46 ± 1.66 ^{aA}	10.76 ± 1.59 ^{abB}	18.45 ± 2.21 ^{aA}	19.58 ± 2.69 ^{aA}
62.5	46.87 ± 1.79 ^{bA}	68.40 ± 1.08 ^{aB}	12.50 ± 0.97 ^{aA}	11.15 ± 0.99 ^{aB}	17.33 ± 1.03 ^{aA}	21.53 ± 2.78 ^{bB}
71.1	47.08 ± 1.57 ^{bA}	66.45 ± 1.85 ^{bB}	12.63 ± 1.28 ^{aA}	10.58 ± 0.64 ^{abB}	17.45 ± 1.04 ^{aA}	19.59 ± 1.84 ^{aB}
76	46.67 ± 1.59 ^{bA}	68.45 ± 1.55 ^{aB}	11.86 ± 0.75 ^{aA}	10.03 ± 0.53 ^{bB}	16.96 ± 1.04 ^{aA}	18.71 ± 2.09 ^{aB}
(B) Internal parts						
55	50.25 ± 7.15 ^{aA}	70.99 ± 3.00 ^{aB}	28.06 ± 4.59 ^{aA}	17.19 ± 2.43 ^{aB}	24.42 ± 2.77 ^{aA}	18.84 ± 1.27 ^{aB}
62.5	54.20 ± 1.85 ^{bA}	70.09 ± 3.32 ^{aB}	21.76 ± 5.76 ^{bA}	16.25 ± 3.60 ^{aB}	21.31 ± 2.66 ^{bA}	18.49 ± 1.79 ^{aB}
71.1	53.79 ± 2.35 ^{bA}	72.97 ± 2.64 ^{aB}	19.34 ± 6.72 ^{cA}	12.20 ± 1.15 ^{bB}	20.45 ± 2.97 ^{cA}	16.26 ± 0.59 ^{bB}
76	53.66 ± 1.74 ^{bA}	72.35 ± 2.01 ^{aB}	13.95 ± 2.63 ^{dA}	11.21 ± 0.97 ^{bB}	17.94 ± 1.49 ^{dA}	15.57 ± 0.32 ^{bB}

Mean values with a different lowercase letter within a column within L*, a*, and b*are significantly different ($p < 0.05$); mean values with a different capital letter within a row within L*, a*, and b* are significantly different ($p < 0.05$).

Limited studies reported an internal color variation in ground veal when cooked to different end-point temperatures. In this study, a similar color variation tendency was detected in veal samples compared to that in beef samples. In cooked veal patties, cooking to an end-point temperature of 71.1 °C or 76 °C resulted in a lower ($p < 0.05$) a* value of 11.21 to 12.2 and a lower b* value of 15.56 to 16.25 than those of the samples that were cooked to 55 °C or 62.5 °C, with an a* value of 16.25 to 17.19 and a b* value of 18.49 to 18.84 (Table 4(B)). However, there was no difference ($p > 0.05$) in L* values, ranging from 70.09 to 72.96, among the veal samples that were cooked from 55 °C to 76 °C (Table 4(B)). Cooked color is important to as consumers use it in determining degree of doneness when consuming ground beef and veal. Using color exclusively could lead to the consumption of undercooked ground beef and veal, therefore, increasing the risk of foodborne illness from pathogenic bacteria.

3.4. Survival Curves of *E. coli* O157:H7 in Course Ground Beef and Veal Patties

Data points shown in the Figure 2 illustrate the survival curves of *E. coli* O157:H7 in non-intact course ground beef and veal patties after cooking at 177 °C (or 350 °F) with various heating times. As expected, the pathogen population in beef and veal samples decreased with the increasing of heating time. After cooking at 177 °C for 360 s, a reduction of 5.67 and 6.42 log CFU/g was observed in ground beef and veal samples, respectively (Figure 2). For both beef and veal samples, it was noticed that the pathogen population did not decrease significantly at the early stage (less than 120–180 s) of cooking, but when the heating time exceeded 180 s the rate of reduction started to accelerate (Figure 2). These results can be explained by the "shoulder effect" [6], which suggested that the thermal inactivation affected by the dimension of the beef and veal patties causing the geometric center temperature did not increase immediately and the pathogen located at the geometric center were not killed at the early stage.

In this study, three survival models in the USDA-IPMP software were used to evaluate the fitness of the model to predict the thermal inactivation kinetics of *E. coli* O157:H7 cells in beef and veal samples (i.e., low value of RMSE and AIC). As shown in Table 5, the Mafart-Weibull and Buchanan Two-Phase Linear models are equally fit for describing the thermal kinetics data of beef and veal samples based on their lower RMSE (0.212 to 0.223 for beef and 0.436 to 0.516 for veal) and lower AIC scores (−67.706 to −65.257 for beef and −24.960 to −33.039 for veal) compared to the Linear model. The similar α value of beef (3.45) and veal (2.77) samples obtained from the Mafart-Weibull model indicated that the pathogen survival curves of beef and veal are in the same shape and exist an obvious "shoulder" effect [6,25] (Figure 2).

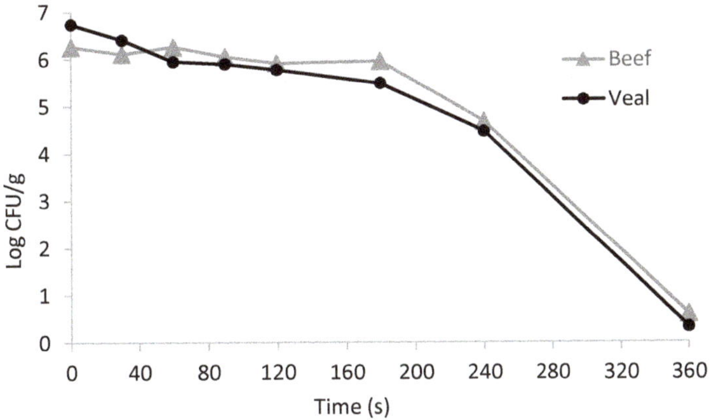

Figure 2. Survival curves of *Escherichia coli* O157:H7 in non-intact course ground beef and veal patties that were cooked by double pan-broiling using a Farberware grill set at 177 °C (or 350 °F).

Previous studies [26,27] has reported the D-value of *E. coli* O157:H7 in ground beef, turkey, lamb, and pork meat, with cooked temperatures from 55 °C to 65 °C in water bath settings, and no studies have reported thermal dynamic parameters of veal products yet. Results of this study showed that the D-value of *E. coli* O157:H7 in ground veal samples cooked at 177 °C on a commercial double pan-broiling griller were 63.27, 189.5, and 29.41 s calculated from Linear, Mafart-Weibull, and Buchanan Two-Phase Linear models, respectively, which were significantly lower than those from the beef samples (Table 6). According to the Buchanan Two-Phase Linear model, the shoulder time of veal is significantly lower than that of the beef samples (167.33 vs 198.19 s, Table 6). These results indicated that *E. coli* O157:H7 cells in veal samples were more sensitive to heat compared to the beef samples.

Table 5. Comparison of square root of mean of sum of squared errors (RMSE) and Akaike information criterion (AIC) value for the proposed survival models on the inactivation of *E. coli* O157:H7 in ground beef and veal patties after double pan-broiling at 177 °C (350 °F) for 0 to 360 s.

Products	Index	Linear	Mafart-Weibull	Buchanan Two-Phase Linear
Beef	RMSE	0.936	0.223	0.212
	AIC	1.752	−65.257	−67.706
Veal	RMSE	0.850	0.436	0.516
	AIC	−2.908	−33.039	−24.960

Table 6. Parameters (mean ± standard error) of survival models estimated for the inactivation of *E. coli* O157:H7 in ground beef and veal patties after double pan-broiling at 177 °C (350 °F) for 0 to 360 s.

Model Name	Model Parameters	Beef	Veal
Linear	D	70.67 ± 8.50 [a]	63.27 ± 6.19 [b]
Mafart-Weibull	D	218.3 ± 7.65 [a]	189.5 ± 14.41 [b]
	α	3.45 ± 0.23 [a]	2.77 ± 0.32 [a]
Buchanan Two-Phase Linear	Shoulder	198.19 ± 5.22 [a]	167.33 ± 10.33 [b]
	D	33.91 ± 2.64 [a]	29.41 ± 1.25 [b]

Means with a different lowercase letters within a row are significantly different ($p < 0.05$).

3.5. Cooking Inactivation of *E. coli O157:H7* Populations with Various Target Temperature and Rest Time

Before cooking, the initial *E. coli* O157:H7 population in uncooked coarse ground beef and veal samples was ranged from 6.4 to 6.6 log CFU/g (Table 7). In general, the total bacterial population counts on TSAP were similar to those that were observed on MacConkey agar in the majority of treatments, indicating that the major colonies that were found on TSAP were *E. coli* O157:H7. However, the recovery of bacterial populations on MacConkey agar was lower than that on TSAP when the veal samples were cooked to 55 °C with a 3.5-min rest and cooked to 62.5 °C with 0.5 min rest. This can be explained by the heat-injured *E. coli* O157:H7 cells not being recovered on MacConkey agar as a selective medium.

Very limited studies have reported the thermal inactivation of *E. coli* O157:H7 in non-intact veal products. Luchansky and others [11] recently found that cooking breaded or un-breaded veal cutlets for 1.5 min per side on an electronic skillet at 191.5 °C achieved an internal temperature of 71.1 °C and a >5.0 log reduction. In a recent study of the same research group, the authors reported that cooking breaded veal cordon bleu at 191.5 °C in pre-heated extra virgin olive oil for ≤6 min or for seven to 10 min per side achieved 1.5 or 3.5 log CFU/g and ≥6.2 log CFU/g, respectively [14]. In this study, we compared the thermal-sensitive *E. coli* O157:H7 in non-intact course ground beef and veal patties. The results of this study indicate that in both beef and veal samples, the higher the cooked internal temperature, the longer the rest time, the greater reductions of *E. coli* O157:H7 was reached, and *E. coli* O157:H7 cells were more ($p < 0.05$) sensitive to heat in veal samples than in beef samples.

As expected, double pan-broiling beef and veal samples to 71.1 °C (well done doneness) and 76 °C (beyond well-done doneness) decreased the overall pathogen populations to below the detection limit (>6 log CFU/g reduction), regardless of the rest time (Table 7), in agreement with the study [28], who reported that cooking refrigerated ground beef patties to internal temperature of 71.1 °C and 76.6 °C reduced *E. coli* O157:H7 to 5.1–7.0 log CFU/g. To reduce the possibility of food-borne outbreaks due to *E. coli* O157:H7 contamination, the USDA-FSIS recommends cooking non-intact veal products to an internal temperature of 62.5 °C (145 °F) with at least a three-minute rest time [29]. To the best of our knowledge, no research publication has reported the impact of rest time on the thermal inactivation activity of *E. coli* O157:H7 on non-intact beef and veal products. By double pan-broiling to 55 °C and 62.5 °C with a 0.5-min rest, 1.95 to 4.79 log CFU/g and 1.97 to >6.0 log CFU/g reductions of *E. coli* O157:H7 were achieved in beef and veal samples, respectively (Table 7). For beef samples, an additional ($p < 0.05$) 0.98 to 1.14 log CFU/g reduction was reached when the rest time extended from 0.5 min to 3.5 min after cooking from 55 °C to 62.5 °C (Table 7). This enhancement of thermal inactivation with a longer rest time was mainly due to the extended heating of beef and veal, causing the internal temperature to continue to increase, even after the patties were removed from the grill. Similar to beef samples, when course ground veal patties were cooked to 55 °C with a 3.5-min rest, an additional ($p < 0.05$) reduction of 2.68 log CFU/g was achieved compared to the 0.5-min rest (Table 7). It is interesting to note that the amount of surviving *E. coli* O157:H7 was below the detect limit (>6.0 log CFU/g reduction) when veal samples were cooked to 62.5 °C with a 0.5- or 3.5-min rest (Table 7), significantly ($p < 0.05$) lower than the same amount in beef samples on MacConkey agar. This result might be explained by the higher moisture content (slightly less fat) in combination with less mature and less thick collagen in veal muscle tissue, allowing for more efficient heat transfer. The compaction density of the physiologically less mature veal tissue versus beef tissue in the patty could also be a factor affecting heat transfer, which produces a greater steaming effect inside the veal patties, resulting greater pathogen reduction. Over all, the present study demonstrates that cooking coarse ground beef and veal patties to an internal end-point temperature of ≥62.5 °C with at least a 3-min rest achieves a >5.0 log reduction of *E. coli* O157:H7 cells.

Table 7. Total bacterial and *Escherichia coli* O157:H7 populations (log CFU/g; ±standard deviation) that were recovered from tryptic soy agar plus 0.1% sodium pyruvate (TSAP) and MacConkey agar, respectively, in beef and veal samples before and after double pan-broiling to 55, 62.5, 71.1, and 76 °C with a 0.5- or 3.5-min rest.

Temperature (°C)	Rest Time (min)	TSAP		MacConkey Agar	
		Beef	Veal	Beef	Veal
After inoculation	-	6.44 ± 0.05 ^{aA}	6.60 ± 0.01 ^{aA}	6.40 ± 0.07 ^{aA}	6.40 ± 0.09 ^{aA}
Before cooking	-	6.37 ± 0.65 ^{aA}	6.58 ± 0.06 ^{aA}	6.47 ± 0.33 ^{aA}	6.35 ± 0.16 ^{aA}
55	0.5	4.86 ± 0.08 ^{bA}	4.32 ± 0.24 ^{bA}	4.52 ± 0.29 ^{bA}	3.38 ± 1.10 ^{bB}
	3.5	3.32 ± 0.81 ^{cA}	1.58 ± 1.28 ^{cB}	3.38 ± 1.10 ^{cA}	0.70 ± 0.59 ^{cB}
62.5	0.5	1.87 ± 0.60 ^{dA}	1.46 ± 0.54 ^{cA}	1.68 ± 0.37 ^{dA}	<0.3 ^{dB}
	3.5	0.94 ± 0.58 ^{eA}	<0.3 ^{dB}	0.70 ± 0.59 ^{eA}	<0.3 ^{dB}
71.1	0.5	<0.3 ^{fA}	<0.3 ^{dA}	<0.3 ^{fA}	<0.3 ^{dA}
	3.5	<0.3 ^{fA}	<0.3 ^{dA}	<0.3 ^{fA}	<0.3 ^{dA}
76	0.5	<0.3 ^{fA}	<0.3 ^{dA}	<0.3 ^{fA}	<0.3 ^{dA}
	3.5	<0.3 ^{fA}	<0.3 ^{dA}	<0.3 ^{fA}	<0.3 ^{dA}

Means with a different lowercase letter within a column or a different capital letter within a row on TSAP or MacConkey agar are significantly different ($p < 0.05$).

4. Conclusions

In conclusion, *E. coli* O157:H7 is more vulnerable in veal compared to the beef during thermal processing. A higher internal temperature and longer rest time cause an increased inactivation of *E. coli* O157:H7, and veal and beef patties present similar tendencies in terms of quality change throughout storage and cooking, supporting our hypothesis. The results of this study cover various aspects of beef and veal quality changes during storage and cooking that will be beneficial for intact and non-intact beef and veal preparation at multiple points, including retail, foodservices, and at home. It was also verified that cooking coarse ground beef or veal to an internal end-point temperature of 62.5 °C with a 3.5-min rest will not generate a great food safety risk. This information will be useful for the USDA-FSIS in developing risk assessments of *E. coli* O157:H7 in non-intact and intact beef and veal products.

Acknowledgments: This work was supported by the Western Kentucky University Honors College Research Scholarship Program and West Virginia University New Faculty Start-up Funding.

Author Contributions: KaWang Li designed and conducted this study, including the quality and microbial pathogen studies, interpreted the results, and drafted the manuscript. Amanda Gipe McKeith designed and conducted the quality control studies, participated in the microbial pathogen studies, interpreted the results, and drafted and revised manuscript. Cangliang Shen generated the idea for this study, designed and conducted the microbial pathogen studies, coordinated the collaboration between different institutions, interpreted the results, and revised and finalized the manuscript. Russell McKeith conducted parts of the quality control studies, in particular the color measurement, and assisted in measuring the fat and moisture contents.

Conflicts of Interest: The authors declare no conflict of interest.

References

1. Meng, J.H.; Lejeune, J.T.; Zhao, T.; Doyle, M.P. *Escherichia coli* O157:H7. In *Food Microbiology: Fundamentals and Frontiers*, 1st ed.; Doyle, M.P., Beuchat, L.R., Montville, T.J., Eds.; ASM Press: Washington, DC, USA, 1997; pp. 171–191, ISBN 978-1-55581-117-4.
2. Beef Products Contaminated with *Escherichia coli* O157:H7. Available online: https://www.federalregister.gov/documents/1999/01/19/99-1123/beef-products-contaminated-with-escherichia-coli-o157h7 (accessed on 15 October 2017).
3. USDA-FSIS. Risk Assessment of the Public Health Impact of *Escherichia coli* O157:H7 in Ground Beef. Available online: https://www.fsis.usda.gov/OPPDE/rdad/FRPubs/00-023N/00-023NReport.pdf (accessed on 15 October 2017).
4. Shen, C.; Geornaras, I.; Belk, K.E.; Smith, G.C.; Sofos, J.N. Inactivation of *Escherichia coli* O157:H7 in nonintact beefsteaks of different thickness cooked by pan broiling, double pan broiling, or roasting by using five types of cooking appliances. *J. Food Prot.* **2010**, *73*, 461–469. [CrossRef] [PubMed]
5. Schmidt, T.B.; Keene, M.P.; Lorenzen, C.L. Improving Consumer Satisfaction of Beef Through the Use of Thermometers and Consumer Education by Wait Staff. *J. Food Sci.* **2002**, *67*, 3190–3193. [CrossRef]
6. Huang, L. Thermal inactivation of Listeria monocytogenes in ground beef under isothermal and dynamic temperature conditions. *J. Food Eng.* **2009**, *90*, 380–387. [CrossRef]
7. Shen, C.; Geornaras, I.; Belk, K.E.; Smith, G.C.; Sofos, J.N. Inactivation of *Escherichia coli* O157:H7 in Moisture-Enhanced Nonintact Beef by Pan-Broiling or Roasting with Various Cooking Appliances Set at Different Temperatures. *J. Food Sci.* **2011**, *76*, M64–M71. [CrossRef] [PubMed]
8. Shen, C.; Geornaras, I.; Belk, K.E.; Smith, G.C.; Sofos, J.N. Thermal Inactivation of Acid, Cold, Heat, Starvation, and Desiccation Stress-Adapted *Escherichia coli* O157:H7 in Moisture-Enhanced Nonintact Beef. *J. Food Prot.* **2011**, *74*, 531–538. [CrossRef] [PubMed]
9. National Cattleman's Beef Association. Beef Bytes Veal Trends. Available online: https://www.beefboard.org/uDocs/d_beef_bytes.pdf (accessed on 14 October 2017).
10. Veal Industry in the United States. Available online: http://www.beeffoodservice.com/CMDocs/BFS/BeefU/2011%20Beef%20U/14%20FS%20Veal.pdf (accessed on 18 November 2017).
11. Luchansky, J.B.; Porto-Fett, A.C.S.; Shoyer, B.A.; Thippareddi, H.; Amaya, J.R.; Lemler, M. Thermal Inactivation of *Escherichia coli* O157:H7 and Non-O157 Shiga Toxin-Producing *Escherichia coli* Cells in Mechanically Tenderized Veal. *J. Food Prot.* **2014**, *77*, 1201–1206. [CrossRef] [PubMed]
12. USDA-FSIS. Marcho Farms, Inc. Recalls Veal, Beef and Pork Products due to Possible Non-O157 Shiga Toxin-Producing *E. coli* (STEC) Adulteration. Available online: https://www.fsis.usda.gov/wps/portal/fsis/topics/recalls-and-public-health-alerts/recall-case-archive/archive/2017/recall-044-2017-release (accessed on 18 November 2017).

13. USDA-FSIS. Annual Report for Shiga Toxin-producing *E. coli* (STEC) in Raw Ground Beef, Veal, and Raw Ground Beef or Veal Components—Calendar Year 2013. Available online: https://www.fsis.usda.gov/wps/wcm/connect/fsis-content/internet/main/topics/data-collection-and-reports/microbiology/ec/stec-annual-report (accessed on 18 November 2017).
14. Kulas, M.; Porto-Fett, A.C.S.; Swartz, R.S.; Shane, L.E.; Strasser, H.; Munson, M.; Shoyer, B.A.; Luchansky, J.B. Thermal Inactivation of Shiga Toxin-Producing *Escherichia coli* Cells within Veal Cordon Bleu: Inactivation of STEC in Cordon Bleu. *J. Food Saf.* **2015**, *35*, 403–409. [CrossRef]
15. Cornforth, D.P.; Jayasingh, P. Chemical and physical characteristics of meat/Color and Pigment. In *Encyclopedia of Meat Sciences*; Werner, K.J., Carrick, D., Michael, D., Eds.; Elsevier Science Ltd.: Oxford, UK, 2004; Volume 1, pp. 249–256. ISBN 978-0-12464-970-5.
16. Shen, C.; McKeith, A.; Broyles, C.; McKeith, R. Quality attributes and thermal inactivation of Shiga toxin-producing *Escherichia coli* in moisture enhanced, non-intact beef products are affected by pump rate, internal temperature, and resting time after cooking. *Food Control* **2016**, *68*, 112–117. [CrossRef]
17. Huang, L. IPMP 2013—A comprehensive data analysis tool for predictive microbiology. *Int. J. Food Microbiol.* **2014**, *171*, 100–107. [CrossRef] [PubMed]
18. Berry, B.W. Cooked Color in High pH Beef Patties as Related to Fat Content and Cooking from the Frozen or Thawed State. *J. Food Sci.* **1998**, *63*, 797–800. [CrossRef]
19. Troutt, E.S.; Hunt, M.C.; Johnson, D.E.; Claus, J.R.; Kastner, C.L.; Kropf, D.H.; Stroda, S. Chemical, Physical, and Sensory Characterization of Ground Beef Containing 5 to 30 Percent Fat. *J. Food Sci.* **1992**, *57*, 25–29. [CrossRef]
20. Lawrie, R.A. The conversion of muscle to meat. In *Lawrie's Meat Science*; Woodhead Publishing in Food Science, Technology and Nutrition; CRC Press: Boca Raton, FL, USA, 2006; pp. 128–155, ISBN 978-1-84569-159-2.
21. Yoon, Y.; Mukherjee, A.; Geornaras, I.; Belk, K.E.; Scanga, J.A.; Smith, G.C.; Sofos, J.N. Inactivation of *Escherichia coli* O157:H7 during cooking of non-intact beef treated with tenderization/marination and flavoring ingredients. *Food Control* **2011**, *22*, 1859–1864. [CrossRef]
22. Madhavi, D.L.; Carpenter, C.E. Aging and Processing Affect Color, Metmyoglobin Reductase and Oxygen Consumption of Beef Muscles. *J. Food Sci.* **1993**, *58*, 939–942. [CrossRef]
23. Hague, M.A.; Warren, K.E.; Hunt, M.C.; Kropf, D.H.; Kastner, C.L.; Stroda, S.L.; Johnson, D.E. Endpoint temperature, internal cooked color, and expressible juice color relationships in ground beef patties. *J. Food Sci.* **1994**, *59*, 465–470. [CrossRef]
24. Hunt, M.C.; Sorheim, O.; Slinde, E. Color and Heat Denaturation of Myoglobin Forms in Ground Beef. *J. Food Sci.* **1999**, *64*, 847–851. [CrossRef]
25. Albert, I.; Mafart, P. A modified Weibull model for bacterial inactivation. *Int. J. Food Microbiol.* **2005**, *100*, 197–211. [CrossRef] [PubMed]
26. Juneja, V.K.; Snyder, O.P.; Marmer, B.S. Thermal destruction of *Escherichia coli* O157:H7 in beef and chicken: Determination of D- and z-values. *Int. J. Food Microbiol.* **1997**, *35*, 231–237. [CrossRef]
27. Juneja, V.K.; Marmer, B.S. Lethality of heat to *Escherichia coli* O157:H7: D- and z-value determinations in turkey, lamb and pork. *Food Res. Int.* **1999**, *32*, 23–28. [CrossRef]
28. Luchansky, J.B.; Porto-Fett, A.C.S.; Shoyer, B.A.; Phillips, J.; Chen, V.; Eblen, D.R.; Cook, L.V.; Mohr, T.B.; Esteban, E.; Bauer, N. Fate of Shiga Toxin-Producing O157:H7 and Non-O157:H7 *Escherichia coli* Cells within Refrigerated, Frozen, or Frozen Then Thawed Ground Beef Patties Cooked on a Commercial Open-Flame Gas or a Clamshell Electric Grill. *J. Food Prot.* **2013**, *76*, 1500–1512. [CrossRef] [PubMed]
29. USDA-FSIS. Veal from Farm to Table. Available online: https://www.fsis.usda.gov/wps/portal/fsis/topics/food-safety-education/get-answers/food-safety-fact-sheets/meat-preparation/veal-from-farm-to-table/ct_index (accessed on 15 October 2017).

Article

Effect of Par Frying on Composition and Texture of Breaded and Battered Catfish

Peter J. Bechtel [1,*], John M. Bland [1], Kristin Woods [2], Jeanne M. Lea [1], Suzanne S. Brashear [1], Stephen M. Boue [1], Kim W. Daigle [1] and Karen L. Bett-Garber [1]

1 Southern Regional Research Center, Agricultural Research Service, United States Department of Agriculture, 1100 Robert E. Lee Blvd., New Orleans, LA 70124, USA; john.bland@ars.usda.gov (J.M.B.); jeanne.lea@ars.usda.gov (J.M.L.); suzanne.brashear@ars.usda.gov (S.S.B.); steve.boue@ars.usda.gov (S.M.B.); kim.daigle@ars.usda.gov (K.W.D.); karen.bett@ars.usda.gov (K.L.B.-G.)

2 Alabama Cooperative Extension System, Auburn University, 120 Court Street, Grove Hill, AL 36451, USA; woodskl@aces.edu

* Correspondence: peter.bechtel@ars.usda.gov; Tel.: +1-504-286-4448

Received: 12 February 2018; Accepted: 19 March 2018; Published: 23 March 2018

Abstract: Catfish is often consumed as a breaded and battered fried product; however, there is increasing interest in breaded and battered baked products as a healthier alternative. Par frying can improve the texture properties of breaded and battered baked products, but there are concerns about the increase in lipid uptake from par frying. The objective of this study was to examine the effect of different batters (rice, corn, and wheat) and the effect of par frying on the composition and texture properties of baked catfish. Catfish fillets were cut strips and then coated with batters, which had similar viscosities. Half of the strips were par fried in 177 °C vegetable oil for 1 min and the other half were not par fried. Samples were baked at 177 °C for 25 min. Analysis included % batter adhesion, cooking loss, protein, lipid, ash, and moisture, plus hardness and fracture quality measured using a texture analyzer. A trained sensory panel evaluated both breading and flesh texture attributes. Results found the lipid content of par fried treatments were significantly higher for both corn and wheat batters than for non-par fried treatments. Sensory analysis indicated that the texture of the coatings in the par fried treatments were significantly greater for hardness attributes. Fillet flakiness was significantly greater in the par fried treatments and corn-based batters had moister fillet strips compared to the wheat flour batters. Texture analyzer hardness values were higher for the par fried treatments.

Keywords: catfish; batters; texture; oil content

1. Introduction

Catfish production is the largest US aquaculture industry; and approximately 161 million pounds of catfish were processed in 2012 [1]. The catfish industry produces a large number of breaded and battered catfish products for deep fat frying; however, few breaded and battered products are designed for baking. The most common problem with baked battered and breaded fish products is that the texture is different (less crisp) than the fried product. One method for improving the texture properties of baked products is to par fry prior to baking [2]. An initial par frying step sets the batter and breading and results in an acceptable baked product in terms of texture. Issues associated with par frying fish products are equipment and operation expenses, and the increased percentage of calories from the oil that is adsorbed by the product. A baked product that is not par fried, but that has texture properties approaching those of a fried product, would have the potential to meet purchasing specifications for food service operations, such as school districts, health care facilities and government purchases.

The basic operations involved in catfish processing, and further processing, have been summarized previously [3] and include batter and breading of catfish fillets. There are a number of relevant studies on the batter and breading of fish, including reports on the rheology and other physical properties of batters used with frying products [4–8]. Batters and breading that would alter fat uptake from frying were examined by Ang in 1993 [9], and, more recently, in a study of rice and wheat batters [10]. The texture properties of microwave-cooked breaded and battered fish were evaluated and effects of batter formulation on quality and crispness of studied microwave-cooked fish nuggets [11–13]. The performance effects of different frying oils on the performance used in frying of chicken and fish sticks have been reported [14]. Studies on batter ingredients to reduce oil uptake and other properties have been reported [15–19]. The effects of coating porosity on oil uptake have been reported [20], and the effects of different par frying temperatures on the physical characteristics of silver carp nuggets were examined [21]. Others have evaluated cooking methods on the properties of breaded black ponfret fillets and lipid oxidation of salmon and mackerel fish nuggets during frozen storage [22,23].

There are numerous examples of breaded and battered catfish products available in the market place for deep fat frying. Most battered and breaded fish products on the market, which would be baked, have undergone an initial par frying step that sets the batter and breading and results in an acceptable baked product, in terms of its texture and product appeal, when baked. The overall goal of this study was to make a battered catfish product that could be baked and have a lower percentage of oil-based calories than equivalent par fried products. The specific objectives of this study were to compare the textures and physicochemical parameters of catfish fillet strips battered with rice, corn, and wheat based flours, which would be either baked or par fried and baked.

2. Materials and Methods

2.1. Sample Preparation

Frozen Individual Quick Frozen (IQF) catfish shank fillets (7–9 oz) were obtained from a large catfish processor. The fillets were thawed and cut into 25 to 30 g strips and then coated with either rice flour (Organic White Rice Flour, Arrowhead Mills Inc., Boulder, CO, USA), corn flour (Whole Grain Corn Flour, Bobs Red Mill Natural Foods, Milwaukie, OR, USA), or wheat flour (Unbleached All Purpose Flour, King Author Flour, Norwich, VT, USA) batters. All batters were adjusted to a similar viscosity of around 120 Rapid Visco Units (1440 Cps) by addition of water to the dry mix. The batter formulae follow: Rice flour 95.28 g, xanthan gum 0.1 g, NaCl 3 g, baking powder 1.72 g, and water; corn flour 95.28 g, NaCl 3 g, baking powder 1.72 g, and water; and wheat flour 95.28 g, NaCl 3 g, baking powder 1.72 g, and water. Individual catfish strips were coated with batter and placed on elevated racks. The dimensions of the battered and breaded catfish strips were approximately 10.1 cm in length, 2.7 cm in width and 2.1 cm in thickness. The percent batter adhering to the raw strips was determined from the weight of the individual strips before and after battering. After coating, half of the strips were par fried and the other half were not par fried. Coated strips and par fried coated strips were placed individually on trays in a −20 °C freezer, and then stored in plastic bags at −20 °C until evaluated.

The par fried treatment consisted of frying coated strips in vegetable oil at 177 °C for 1 min, which fell within the parameters of par frying [2]. After the par frying step, the strips were placed on paper towels until cool and placed in a −20 °C freezer. The baking treatment for both par fried and non-par-fried samples was 177 °C for 25 min. Cooking loss was determined for the par fried treatment by weighing battered and breaded strips before and after par frying, but before freezing. Cooking loss was determined for baked samples by weighing strips before and after baking.

2.2. Proximate Analysis

Moisture and ash content were determined using Association of Official Analytical Chemists (AOAC) (1990) methods #950.46 and #923.03 [24], respectively. Nitrogen content was determined using pyrolysis with a Leco TruSpec N nitrogen analyzer (Leco Co., St. Joseph, MO, USA). Protein content was calculated as 6.25 times percent N. Moisture, ash, and lipid contents were determined in duplicate and protein content were determined either in duplicate or triplicate for each replicate sample. Total lipid content was determined gravimetrically using a modification of the Folch procedure [25], using a Dionex ASE 200 accelerated solvent extractor (Dionex (UK) Ltd., Camberley, Surrey, UK) with methylene chloride at 100 °C and 1500 psi. After lipid extraction, the solvent was removed under a N_2 gas stream at 40 °C using a TurboVap LV (Caliper Life Sciences, Waltham, MA, USA) in pre-weighed vials [26].

2.3. Color Measurement

The color of battered fillet strips were measured after baking using a Konica-Minolta Chroma meter CR-410 (Konica-Minolta Sensing, Tokyo, Japan). The colorimeter was calibrated with a white C standard plate ($Y = 86.1$, $x = 0.3154$ and $y = 0.3229$) and 2° observer angle before sample measurements. CIE Lab color space was used to record L^*, a^* and b^* values. The L^* represents the lightness of the color, where 100 represents white and zero represents black. The a^* axis indicates a red shade when greater than zero (positive) and a green shade when lower than zero (negative). The b^* axis indicates a yellow shade when positive and a blue shade when negative. Color values were determined for multiple locations on a strip and then an averaged value was determined for each strip.

2.4. Sensory Evaluation

The breading texture attributes for this experiment were selected from those previously reported [10,27]. The attributes selected were flaky, hardness, fracturability, crispness and tooth packing [28], and are described in Table 1. The cooked fillet texture attributes used were adapted from Chambers and Robel [29] and included flaky, firmness, moisture release, fibrous, moisture retention, and cohesiveness of mass (Table 2).

Table 1. Breading texture attribute lexicon.

Attribute	Definition and Reference
Hardness	The force to compress the food. 1.0 = Philadelphia Light Cream Cheese to 11.0 Uncooked carrots
Fracturability	The force with which the sample crumbles or shatters. 1.0 = Jiffy Corn Muffin to 13.0 Peanut Brittle
Crispness	The force and noise with which a product breaks rather than deforms. 3.0 = Quaker Chewy Granola Bar to 17.0 Old London Melba Toast
Toothpacking	Degree to which product sticks to the surface of the teeth. 1.0 = Raw carrots to 11.0 Cheetos

Ten panelists were trained to rate intensities of texture attributes for battered and breaded catfish. They practiced for six sessions using fish prepared in the following way: approximately 30 g of fish was fried in 177 °C vegetable oil for 3.5 min, par-fried in 177 °C oil for 1 min and then baked at 218 °C for 15 min, or baked at 218 °C for 25 min. During practice sessions, Zatarain's Crispy Southern fish seasoning, New Orleans, LA, USA (with corn meal), Zatarain's chicken frying mix, New Orleans, LA (with wheat flour), and Choice Batter (by CrispTek, LLC, Columbia, MD, USA) (with rice flour), were used to represent experimental formulations. One session was used to practice on breaded par-fried and breaded-only samples that were pre-frozen. The frozen samples were baked prior to presentation to panelists for evaluation. Prepared fish samples were served in warmed glass bowls placed in foam bowls to keep them warm. Samples were presented monadically to the panelists under sodium vapor lights that mask color variation. Sensory evaluations were conducted in individual booths within a climate controlled room, and scores were entered into Compusense five 5.4.15v (Guelph, ON, Canada) sensory balloting software. Unsalted soda crackers and filtered water were provided to cleanse the palate between each test sample.

Table 2. Catfish fillet flesh sensory texture attributes.

Attribute	Definition and Reference
Flaky-Visual	The ease of breaking the fish into small pieces with a fork. 2.0 = Deli Turkey to 7.0 = Bumble Bee Fancy Lump crab meat
Firmness	Amount of force required to bite through the flesh when the sample is placed between molar teeth. (On hot dog, place cut surfaces between molars.) 4.0 = Hebrew National All beef hot dog to 9.0 = Cooked chicken breast
Moisture Release (Juicy Initial)	Bite with molars then evaluate the amount of liquid released when the sample is placed on tongue and pressed to the roof of the mouth. 2.0 = Oscar Meyer All beef hot dog, 11.0 = sliced orange
Fibrous	The perception of filaments or strands of muscle tissue during mastication. 2.0 = Ball Park All Beef Hot Dog, 10.0 = Cooked chicken breast
Moisture Retention (Juicy Mid-Point)	Amount of liquid observed in the mass after 5 chews with the molar teeth. 4.0 = Deli Turkey, 7.0 = Ball Park All Beef Hot Dog
Cohesiveness of Mass	The degree to which chewed sample (at 10 to 15 chews) holds together in a mass (forms a ball). 3.0 = Raw Mushroom, to 8.0 = Cooked Chicken Breast

Experimental samples for sensory evaluation (see Section 2.1) were battered and individually frozen prior to the panel sessions. The par-fried samples were par-fried prior to freezing. At each experimental sensory evaluation session, two different flour treatments (with one baked sample and one par-fried sample) were presented. Both the baked sample and the par-fried then baked sample were baked at 177 °C for 25 min. Samples were served as described above. Each flour-batter treatment/cooking method combination was presented at two different panel sessions and scores were entered into Compusense five. All three flour types (corn, rice, and wheat) were paired with the other flour-batter types during one session.

2.5. Texture Analysis

Textural properties were measured on a Steven's QTS Texture Analyzer (Brookfield Engineering Labs, Inc., Middleboro, MA, USA) with a 25-kg load cell using a TA-52 stainless steel shear blade. The texture was measured after samples had been baked for 25 min at 177 °C and then cooled at 21 °C for 30 min. The cooked fish was compressed in a single cycle test at 150-mm/min-test speed until a 50% deformation target was reached. Textural properties were calculated using Texture Pro2 software from Brookfield Engineering Labs, Inc. (Middleboro, MA, USA). Recorded properties included hardness, defined as the peak compression force attained in the force deformation curve, the quantity of fractures, defined as the number of occasions the load decreased by 5% prior to reaching the target value, and the sample length.

2.6. Statistical Analysis

The means of data were compared using an analysis of variance (ANOVA) and the mean comparison test, Tukey-Kramer adjustment to Least Squares Means, were performed in Proc Mixed using Enterprise Guide, version 5.1, (SAS, Inc., Cary, NC, USA). Significance was reported at $p < 0.05$ for all data.

3. Results and Discussion

3.1. Cooking Yield and Proximate Analysis

The average weight of the raw catfish strips used for the battering experiment was 27.5 g to 28.5 g (Table 3). After battering, the catfish strips weight increased 39.4% to 23.7%, with corn batter having the largest weight gain and wheat batter the least. A breading and batter adhesion of 25% to 30% would be reasonable for chicken or pork nuggets [2]. In this study, the fillet strips had less uniform shapes than a reformed nugget product. When the batter coated product was baked, there was a loss of weight from 27.1% to 24.5%. Par frying for 1 min at 177 °C resulted in a weight loss of 23.3% to 26.9% for the corn and wheat flour batters, respectively. Meanwhile, rice flour battered fish had a weight loss of only 15.9% in the study by Ojagh et al. [21], and they reported a product yield of approximately 80% after par frying, which was higher than the yields found in this study for the wheat and corn flour batters.

Table 3. Weights of par fried and baked catfish strips coated with corn, wheat and rice batters.

	Corn		Wheat		Rice	
	Weight (g)	SD	Weight (g)	SD	Weight (g)	SD
raw	27.7	1.3	28.3	2.8	28.5	1.6
coated raw	38.6	1.8	35.0	2.8	35.9	1.7
baked	28.8	1.8	25.5	2.0	27.1	1.7
par fried	29.6	2.6	25.6	3.2	30.2	3.0
par fried + baked	24.5	3.1	17.9	1.3	25.7	3.2

Mean values with standard deviations (SD); $n = 30$ raw catfish strips; $n = 15$ raw coated catfish strips; $n = 10$ raw coated catfish strips baked; $n = 15$ raw coated catfish strips par fried; $n = 10$ raw coated catfish strips par fried and baked.

The proximate analytical results for raw, baked, par fried (before baking) and par fried then baked samples are listed in Table 4. Within a batter treatment the par fried then baked treatments for the corn and rice batters showed significantly lower ($p < 0.05$) moisture content than coated raw, baked or par fried samples. However, for the wheat coated treatments the par fried (before baking) moisture value was not different from the coated raw value. In addition, the par fried baked moisture value was significantly different ($p < 0.05$) from the baked value for corn and rice batters. There were significant differences between the ash contents; however, ash values ranged from a low of 1.3% to a high of 2.2% on a wet weight basis. Within the baked treatments, the corn and rice battered treatments yielded significantly higher ($p < 0.05$) in moisture content than wheat baked treatments. The protein content in the par fried then baked fish was significantly higher ($p < 0.05$) than the coated raw fish with corn batter. For the wheat batter, the protein values of both par fried then baked and the baked fish were significantly higher than their raw products. This was due to the loss of moisture during cooking of the fish. Par fried and baked corn flour battered fish was significantly greater ($p < 0.05$) in protein than the raw coated fish, but the baked corn flour coated sample was not significantly different ($p < 0.05$) from the raw coated sample. The corn batter baked fish retained more moisture during baking than the wheat battered fish. The wheat baked product was significantly higher in ($p < 0.05$) protein than the corn or rice battered fish, because wheat flour (~14%) typically has a higher protein content than corn (~9%) and rice (~8%) flours. There were no significant differences ($p < 0.05$) in the protein content between the par fried than baked fish for the corn or wheat batter. The moisture of par fried breaded black pomfret values was reported to be 57%, and, after further baking, the moisture content was 56% [22]. In our study, the moisture content of par frying before baking treatments ranged from 65.8% to 72.3%; however, when the par fried product was baked, the moisture content was further reduced to 56.1% to 61.2%. The moisture and oil content of coated par fried Talang queen fish nuggets were reported to be 54.2% and 7.21%, respectively [18].

Table 4. Composition on a wet weight basis of baked and par fried and then baked catfish strips coated with corn, wheat and rice batters.

		% Moisture		% Ash		% Protein		% Lipids (Oil)	
		Mean	SD	Mean	SD	Mean	SD	Mean	SD
Corn	coated raw	73.2 A	1.4	1.4 C	0.2	13.5 B	1.0	3.3 B	0.2
	baked	66.7 Bz	1.5	2.0 AB	0.1	16.6 ABy	1.1	3.0 By	0.1
	par fried before baking	65.8 B	3.0	1.6 BC	0.1	14.9 AB	2.7	7.7 A	1.8
	par fried after baking	56.1 C	2.5	2.1 A	0.2	18.3 A	1.7	9.8 A	0.1
Wheat	coated raw	77.5 A	1.8	1.3 B	0.0	12.6 B	0.8	3.6 B	0.4
	baked	57.9 By	1.1	2.1 A	0.4	23.0 Az	2.5	6.4 Az	1.0
	par fried before baking	72.3 A	0.4	1.6 AB	0.2	15.1 B	0.7	8.0 A	0.7
	par fried after baking	60.3 B	4.8	2.2 A	0.3	20.1 A	2.5	6.9 A	1.5
Rice	coated raw	73.6 A	1.6	1.4 C	0.1	13.9 *	NA	3.6 B	1.4
	baked	68.2 Bz	2.2	1.8 A	0.1	16.1 y	1.1	2.5 By	0.6
	par fried before baking	70.4 AB	1.5	1.6 B	0.0	16.1	1.2	4.2 B	1.2
	par fried after baking	61.2 C	1.8	1.9 A	0.0	18.7	2.7	10.8 A	2.6

Mean % w/w with standard deviation from analysis of 3 strips except the protein value for rice (*) coated raw which was a value for a single strip and par fried before bake and par fried baked which had values from 2 strips. A, B, C indicate that means (between raw, baked, par fried before bake, and par fried baked within the same batter) with different letters are significantly different ($p < 0.05$). z, y indicate that means (between baked of each batter) with different letters are significantly different ($p < 0.05$). The carbohydrate content of the stripes were not determined.

For the corn, rice and wheat flour batter treatments, the par fried then baked treatments compared to the coated raw treatments yielded significantly higher ($p < 0.05$) percent oil content. For both corn and rice flour batters, the percent oil content was significantly greater ($p < 0.05$) for the par fried then baked treatment compared to the baked treatment. The increase in percent oil between the baked and par fried then baked treatments for the corn and rice batters was 6.8% and 8.3%, respectively. These values for corn and rice batters represent increases of over 200% in the calories derived from oil between baked and par fried then baked treatments. For the wheat batter treatments, the percent oil values were not significantly different ($p < 0.05$) for the baked, par fried before baking, and par fried after baking samples; however, these treatments had significantly greater percent oil values ($p < 0.05$) than the coated raw treatment. The oil content of par fried breaded black pomfret values was reported to be 8.8%, which increased after being oven baked to 9.1% in sunflower oil and 9.5% in palm oil [22]. In our study the par fried before baking treatment had percent oil values ranging from 4.2% for the rice flour batter to 8.0% for the wheat flour batter treatment. After baking the par fried then baked, treatments had percent oil values of 9.8%, 6.9% and 10.8% for corn, wheat and rice, respectively, with rice batter having a significant increase ($p < 0.05$) between par frying and baking after par frying.

One method to avoid increasing the oil content of battered fish products is baking without par frying the product, which is one of the treatments in this study. Another method to reduce the oil content of par fried products is to incorporate ingredients in the breading and battering that will reduce the amount of oil adsorbed on the par fried product [13,15,16,18], such as carboxyl methyl cellulose, HPMC, methylcellulose edible coatings, or whey protein concentrate films.

3.2. Color Properties

Colorimetric values of the par-fried and non-par-fried fish strips are listed in Table 5. For the wheat and corn batters, the L^* value was higher for baked fish than the par fried baked treatment, which indicates that the baked strips had a lighter color. However, for the rice batter, the par fried batter was lighter. Within the par fried treatments, the L^* value for the rice batter was significantly higher ($p < 0.05$) than both the corn and wheat batters. For the baked treatments, the rice and wheat batters yielded significantly higher ($p < 0.05$) L^* values than the corn batter. For the a^* color value, the par fried wheat and corn battered fish exhibited significantly higher ($p < 0.05$) than their baked counterparts, making them more red (Table 5). For par fried fish, the a^* value for the corn batter was significantly higher ($p < 0.05$) than the wheat or rice batter treatments. The corn baked fish was significantly higher ($p < 0.05$) than the rice baked, which was significantly higher than wheat baked. The b^* value for par fried rice was significantly higher ($p < 0.05$) than the baked sample. However, in the baked wheat sample, b^* was higher than par fried; for the corn sample, the b^* values were almost equal. The corn par fried sample is significantly higher ($p < 0.05$) than the wheat or rice par fried treatments confirming an increase in yellowness. For the baked treatments, the corn battered sample was significantly higher ($p < 0.05$) than the wheat battered, which was significantly higher than the rice battered samples.

Table 5. Color analysis catfish strips coated with corn, wheat and rice flour batters.

		L^*		a^*		b^*	
		Mean	SD	Mean	SD	Mean	SD
Corn	baked	60.86 ^{y}	2.67	1.45 ^{Bz}	0.11	34.06 ^{z}	2.75
	par fried & baked	58.15 ^{y}	1.13	4.7 ^{Az}	1.1	33.87 ^{z}	3.07
Wheat	baked	66.24 ^{z}	0.97	−0.15 ^{Bx}	0.18	22.15 ^{y}	0.97
	par fried & baked	62.07 ^{y}	4.96	0.67 ^{Ay}	0.17	17.73 ^{y}	2.87
Rice	baked	68.11 ^{z}	3.13	0.49 ^{Ay}	0.28	9.39 ^{Bx}	0.62
	par fried & baked	71.85 ^{z}	2.24	−0.82 ^{By}	0.07	15.94 ^{Ay}	0.85

Mean values with standard deviation from analysis of 10 strips. A, B indicate that means (between baked and par fried within the same batter) with different letters are significantly different ($p < 0.05$). z, y and x indicate that means (between par fried of each batter and baked of each batter) with different letters are significantly different ($p < 0.05$).

In the study by Moradi et al. [22], they reported a decrease in the L^* value and an increase in the a^* and b^* values when par fried black pomfret samples were baked. Others have reported that, when silver carp nuggets were par fried and the oil temperature increased, the L^* value decreased and a^* and b^* values increased [21]. The ideal color for fried products was reported to be a light brown color [30], which would correspond to L^*. a^*, b^* values of approximately 70, 9 and 5, respectively. The a^* values of the catfish strips for all batters were lower, indicating less red in all samples.

3.3. Sensory Evaluation of Texture

There were significant interactions between cooking method and type of flour used in the breading formulations. The type of coating had a significant difference ($p < 0.05$) on all the breading texture attributes (Table 6). For the baked treatments, rice flour was significantly higher ($p < 0.05$) in hardness of the coating than corn and wheat flour. Hardness had a markedly high interaction between flour type and cooking method. In the par fried treatments, hardness, fracturability, crispness and tooth packing of both the corn and rice flours were significantly higher ($p < 0.05$) than the wheat flour. For tooth packing, the rice flour was significantly higher ($p < 0.05$) than the wheat flour. Greater differences were found in the par-fried treatments than in the baked treatments.

The coatings on the par-fried baked catfish strips were significantly greater ($p < 0.05$) in intensity compared to baked coating for hardness, fracturability and crispness texture attributes (Table 6). These attributes had interaction effects that were driven by the type of flour. Studies on the sensory properties par fried fish products include evaluation of silver carp fish nuggets, par fried at different temperatures [21]. However, they did not observe significant differences ($p < 0.05$) between par frying temperatures.

The only significant difference ($p < 0.05$) between batters in the fish fillet texture was moisture retention, where the corn flour for the par fried sample was significantly higher ($p < 0.05$) than the par fried wheat and rice flour treatments (Table 7). Corn flour retained significantly ($p < 0.05$) more moisture than wheat and rice battered products. For the flaky attribute, the corn par fried sample was significantly higher ($p < 0.05$) than the rice par fried fish. The flaky attribute was significantly greater ($p < 0.05$) in the par-fried then baked fish than in the baked fish for corn batter (Table 7), which indicated that the baked fish was more difficult to flake.

Table 6. Sensory properties for batter attributes.

Batter Attributes		Hardness		Fracturability		Crispness		Tooth Packing	
		Mean	SD	Mean	SD	Mean	SD	Mean	SD
Corn	baked	3.35 ^{By}	1.43	2.42 ^{B}	1.62	3.03 ^{Bz}	1.64	5.09 ^{B}	2.02
	par fried & baked	6.06 ^{Az}	1.34	5.26 ^{Az}	1.26	6.12 ^{Az}	1.05	5.88 ^{Az}	1.66
Wheat	baked	2.94 ^{By}	1.03	2.26	1.83	2.13 ^{By}	1.7	5.28	2.02
	par fried & baked	4.38 ^{Ay}	0.94	2.71 ^{y}	1.58	3.09 ^{Ay}	1.53	4.34 ^{y}	1.68
Rice	baked	4.61 ^{Bz}	1.64	3.06 ^{B}	1.76	3.17 ^{Bz}	1.71	5.47 ^{B}	1.55
	par fried & baked	6.16 ^{Az}	1.52	4.47 ^{Az}	1.3	5.53 ^{Az}	1.54	6.29 ^{Az}	1.43

A, B indicate that means (between par fried and baked of the same batter) with different letters are significantly different ($p < 0.05$); z, y indicate that means (between par fried of each batter and baked of each batter) with different letters are significantly different ($p < 0.05$).

Table 7. Sensory properties for fish flesh texture attributes.

		Flaky		Firmness		Moisture Release		Fibrous		Moisture Retention		Cohesiveness of Mass	
		Mean	SD	Mean	SD	Mean	SD	Mean	SD	Mean	SD	Mean	SD
Corn	baked	3.85 ^{B}	1.46	3.85	1.41	5.03	1.64	4.79	1.41	5.15	1.49	5.35	1.26
	Par fry & Baked	5.35 ^{Az}	1.48	3.79	1.52	5.41	1.53	4.82	1.42	5.65 ^{z}	1.44	5.74	1.58
Wheat	baked	4.28	1.79	4.56	1.91	4.83	1.21	4.69	1.09	4.69	0.93	5.28	1.09
	par fry & baked	4.68 ^{zy}	1.54	3.82	1.07	4.76	1.67	4.34	1.3	4.68 ^{y}	1.53	5.5	1.22
Rice	baked	3.81	1.5	4.51	1.79	4.58	1.63	4.83	1.21	4.72	1.49	5.03	1.02
	par fry & baked	4.28 ^{y}	1.22	4.25	1.92	4.61	1.54	4.89	1.49	4.66 ^{y}	1.36	4.88	1.33

A, B indicate that means (between par fry and baked of the same batter) with different letters are significantly different based on Tukey-Kramer adjustment to the least square mean test. z, y indicate that means (between par fry of each batter and baked of each batter) with different letters are significantly different ($p < 0.05$) based on Tukey-Kramer adjustment to the least square mean test.

3.4. Mechanical Texture Properties

The mechanical texture attribute of hardness was greater in the par fried then baked treatments than the baked treatments (Table 8). The par fried corn and rice coated fish samples had significantly greater ($p < 0.05$) hardness than the baked equivalents. There were no significant differences ($p < 0.05$) in hardness between wheat, corn or rice coated par fried then baked fish treatments. Neither was there a significant difference in hardness among the wheat, corn or rice coated baked fish treatments. There was no significant difference (>0.05) in the number of fractures between par-fried then baked and baked fish for each coating. However, the number of fractures were always higher in the par-fried fish. Within the baked fish samples, quantities of fractures for rice coated fish were significantly higher ($p < 0.05$) than the wheat coated fish. For par-fried then baked fish, the quantity of fractures was significantly higher ($p < 0.05$) in the rice coated fish than the wheat coated fish. Moradi et al. (2009) showed that the texture (adhesiveness, springiness, and cohesiveness) values of par fried black pomfret increased slightly when par fried samples were baked.

Table 8. Mechanical texture properties of catfish strips coated with corn, wheat and rice batters.

		Hardness		Quant Fract	
		Mean	SD	Mean	SD
Corn	baked	203.7 B	30.2	67.6 Azy	10.3
	par fried & baked	255.5 A	66.2	73.3 Azy	10.5
Wheat	baked	182.6 A	77.6	57.2 Ay	15.9
	par fried & baked	254.1 A	87.8	64.3 Ay	13.9
Rice	baked	200.1 B	38.9	71.6 Az	8.6
	par fried & baked	310.3 A	60.4	79.5 Az	14.6

Mean values of 10 strips per treatment with standard deviations. Quant Fract is the quantity of fractures (see Section 2). A, B indicate that means (between par fried then baked and baked of the same batter) with different letters are significantly different ($p < 0.05$). z, y indicate that means (between par fried then baked or each batter and baked for each batter) with different letters are significantly different ($p < 0.05$).

4. Conclusions

The oil content of par fried baked products was significantly higher ($p < 0.05$) for the corn and wheat batters compared to that of comparable non par fried treatments. Sensory results indicated that the coatings on the par fried then baked catfish strips were significantly greater ($p < 0.05$) than baked coating for hardness, facturability and crispness attributes. These attributes had interaction effects that were driven by the type of batter. Fillet flakiness was significantly greater ($p < 0.05$) in the corn battered par fried then baked treatment than in the corn battered baked treatment. The corn flour battered products had moister fillet strips. Hardness values measured with a texture analyzer were higher for the par fried then baked treatment when compared to the non-par fried treatment.

Acknowledgments: We thank Carissa Li for critical review of the manuscript. Intramural funding for this study was provided by United States Department of Agriculture, Agricultural Research Service projects 6054-43440-049-00D and 6054-44000-078-00D. Mention of a trade name, proprietary product or specific equipment does not constitute a guarantee or warranty by the US Department of Agriculture and does not imply approval of the product to the exclusion of others that may be available.

Author Contributions: P.J.B. wrote the paper and contributed to the study design. J.M.B. had the responsibility for making products and data analysis. K.W. worked on both the initial pilot study and the manufacture of product and helped design product for use school lunch programs. S.S.B. analyzed the composition of the product. J.M.L. provided statistical analysis and helped with sensory panels. S.M.B. and K.W.D. provide the mechanical texture analysis of products and viscosity measurements of batters. K.L.B.-G. conducted sensory evaluation of products and data analysis.

Conflicts of Interest: All authors declare no conflicts of interest.

References

1. United States Department of Agriculture (USDA). Aquaculture Data, Catfish and Trout: Inventories, Sales, and Prices. 2014. Available online: https://www.ers.usda.gov/data-products/aquaculture-data.aspx#28153 (accessed on 20 March 2018).
2. Barbut, S. *Producing Battered and Breaded Meat Products*; American Meat Science Association: Champaign, IL, USA, 2012.
3. Silva, J.L.; Nunez, A.; Chamul, R.S. *Onset of Rigor and Freshness Quality of Channel Catfish Fillets as Affected by Preprocessing Handling and Transport*; World Aquatic Society: Orlando, FL, USA, 2001; pp. 21–25.
4. Xue, J.; Ngadi, M. Rheological properties of batter systems formulated using different flour combinations. *J. Food Eng.* **2006**, *77*, 334–341. [CrossRef]
5. Chen, H.H.; Kang, H.Y.; Chen, S.D. The effects of ingredients and water content on the rheological properties of batters and physical properties of crusts in fried foods. *J. Food Eng.* **2008**, *88*, 45–54. [CrossRef]
6. Albert, A.; Varela, P.; Salvador, A.; Fiszman, S.M. Improvement of crunchiness of battered fish nuggets. *Eur. Food Res. Technol.* **2009**, *228*, 923–930. [CrossRef]

7. Dogan, S.F.; Sahin, S.; Sumnu, G. Effects of soy and rice flour addition on batter rheology and quality of deep-fat chicken nuggets. *J. Food Eng.* **2005**, *71*, 127–132. [CrossRef]
8. Cunningham, F.E.; Tiede, L.M. Influence of batter viscosity on breading of chicken drumsticks. *J. Food Sci.* **1981**, *46*, 1950–1952. [CrossRef]
9. Ang, J.F. Reduction of fat in fried batter coatings with powdered cellulose. *J. Am. Oil Chem. Soc.* **1993**, *70*, 619–622. [CrossRef]
10. Shih, F.; Bett-Garber, K.; Champagne, E.; Daigle, K.; Lea, J. Effects of beer-battering on the frying properties of rice and wheat batters and their coated foods. *J. Sci. Food Agric.* **2010**, *90*, 2203–2207. [CrossRef] [PubMed]
11. Lopez-Gavito, L.; Pigott, G.M. Effects of Microwave cooking on textural characteristics of battered and breaded fish products. *J. Microw. Power* **1983**, *18*, 345–353. [CrossRef]
12. Chen, C.L.; Li, P.Y.; Hu, W.H.; Lan, M.H.; Chen, M.J.; Chen, H.H. Using HPMC to improve crust crispness in microwave-reheated battered mackerel nuggets: Water barrier effect of HPMC. *Food Hydrocoll.* **2008**, *22*, 1337–1344. [CrossRef]
13. Chen, S.D.; Chen, H.H.; Chao, Y.C.; Lin, R.S. Effect of batter formula on qualities of deep-fat and microwave fried fish nuggets. *J. Food Eng.* **2009**, *95*, 359–364. [CrossRef]
14. Altunakar, B.; Sahin, S.; Sumnu, G. Functionality of batters containing different starch types for deep-fat frying of chicken nuggets. *Eur. Food Res. Technol.* **2004**, *218*, 318–322. [CrossRef]
15. Ballard, T.S.; Mallikarjunan, P. The effect of edible coatings and pressure frying using nitrogen gas on the quality of breaded fried chicken nuggets. *J. Food Sci.* **2006**, *71*, S259–S264. [CrossRef]
16. Brannan, R.G.; Mah, E.; Schott, M.; Yuan, S.; Casher, K.L.; Myers, A.; Herrick, C. Influence of ingredients that reduce oil absorption during immersion frying of battered and breaded foods. *Eur. J. Lipid Sci. Technol.* **2014**, *116*, 240–254. [CrossRef]
17. Baixauli, R.; Sanz, T.; Salvador, A.; Fiszman, S.M. Effect of the addition of dextrin or dried egg on the rheological and textural properties of batters for fried foods. *Food Hydrocoll.* **2003**, *17*, 305–310. [CrossRef]
18. Jamshidi, A.; Shabanpour, B. The effect of hydroxypropyl methylcellulose (HPMC) gum added to predust and batters of talang queen fish (*Scomberoides commersonnianus*) nuggets on the quality and shelf life during frozen storage (−18 °C). *World J. Fish Mar. Sci.* **2013**, *5*, 382–391.
19. Salvador, A.; Sanz, T.; Fiszman, S.M. Effect of the addition of different ingredients on the characteristics of a batter coating for fried seafood prepared without a pre-frying step. *Food Hydrocoll.* **2005**, *19*, 703–708. [CrossRef]
20. Pinthus, E.J.; Weinberg, P.; Saguy, I.S. Oil uptake in deep fat frying as affected by porosity. *J. Food Sci.* **1995**, *60*, 767–769. [CrossRef]
21. Ojagh, S.M.; Shabanpour, B.; Jamshidi, A. The effect of different pre-fried temperatures on physical and chemical characteristics of silver carp fish (*Hypophthalmichthys molitrix*) nuggets. *World J. Fish Mar. Sci.* **2013**, *5*, 414–420.
22. Moradi, Y.; Bakar, J.; Muhamad, S.S.; Man, Y.C. Effects of different final cooking methods on physico-chemical properties of breaded fish fillets. *Am. J. Food Technol.* **2009**, *4*, 136–145.
23. Tolasa, S.; Lee, C.M.; Calki, S. Lipid oxidation and omega-3 fatty acid retention in salmon and mackerel nuggets during frozen storage. *J. Aquat. Food Prod. Technol.* **2011**, *20*, 172–182. [CrossRef]
24. Association of Official Analytical Chemists (AOAC). *Official Methods of Analysis of the AOAC*, 15th ed.; Methods 950.46 and 923.03; Association of Official Analytical Chemists: Arlington, VA, USA, 1990.
25. Folch, J.; Lees, M.; Sloane-Stanley, G.H. A simple method for the isolation and purification of lipids from animal tissues. *J. Biol. Chem.* **1957**, *226*, 497–509. [PubMed]
26. Li, C.H.; Bland, J.M.; Bechtel, P.J. Effect of Precooking and Polyphosphate Treatment on the Quality of Microwave Cooked Catfish Fillets. *Food Sci. Nutr.* **2017**, *5*, 812–819. [CrossRef] [PubMed]
27. Shih, F.F.; Bett-Garber, K.L.; Daigle, K.W.; Ingram, D. Effects of rice batter on oil uptake and sensory quality of coated fried okra. *J. Food Sci.* **2005**, *70*, S18–S21. [CrossRef]
28. Meilgaard, M.; Carr, B.T.; Civille, G.V. *Sensory Evaluation Techniques*, 4th ed.; CRC Press: Boca Raton, FL, USA, 2007.

29. Chambers, I.V.E.; Robel, A. Sensory characteristics of selected species of freshwater fish in retail distribution. *J. Food Sci.* **1993**, *58*, 508–512.
30. Sanz, T.; Salvador, A.; Velez, G.; Munoz, J.; Fiszman, S.M. Influence of ingredients on the thermo-rheological behavior of batter containing methylcellulose. *Food Hydrocoll.* **2005**, *19*, 869–877. [CrossRef]

Article

Meat Quality Derived from High Inclusion of a Micro-Alga or Insect Meal as an Alternative Protein Source in Poultry Diets: A Pilot Study

Brianne A. Altmann [1], Carmen Neumann [2], Susanne Velten [2], Frank Liebert [2] and Daniel Mörlein [2,*]

1 Department of Animal Sciences, Division Animal Product Quality, University of Göttingen, 37075 Göttingen, Germany; brianne.altmann@agr.uni-goettingen.de
2 Department of Animal Sciences, Division Animal Nutrition Physiology, University of Göttingen, 37077 Göttingen, Germany; carmen.neumann@agr.uni-goettingen.de (C.N.); susanne.velten@agr.uni-goettingen.de (S.V.); flieber@gwdg.de (F.L.)
* Correspondence: daniel.moerlein@agr.uni-goettingen.de; Tel.: +49-551-395611

Received: 23 January 2018; Accepted: 7 March 2018; Published: 8 March 2018

Abstract: The effects on meat quality resulting from alternative dietary protein sources (Spirulina and Hermetia meal) in poultry diets are studied to determine the overall suitability of these ingredients considering state-of-the-art packaging practices—highly oxygenated modified atmosphere packaging (HiOx MAP). We monitored standard slaughterhouse parameters, such as live weight, carcass weight, dressed yield, and pH at 20 min and 24 h post mortem. In addition, we studied the effects that 3 and 7-day storage in HiOx MAP has on the overall product physico-chemical and sensory properties. In addition to previously supported effects of HiOx MAP, we found that meat quality could be improved when Spirulina replaces 50% of the soy protein in broiler diets; however, this substitution results in a dark reddish-yellowish meat colour. On the other hand, the substitution with Hermetia larval meal results in a product that does not differ from the standard fed control group, with the exception that the breast filet has a more intense flavour that decreases over storage time. All-in-all Spirulina and Hermetia meal have the potential to replace soybean meal in broiler diets without deteriorating meat quality.

Keywords: broiler; chicken; breast meat; sensory analysis; modified atmosphere packaging; Spirulina; black soldier fly; *Hermetia illucens*; *M. pectoralis superficialis*

---

1. Introduction

Continued population growth and increasing income levels are driving up the demand for animal-based products and, in turn, is increasing the demand for animal feed resources [1]. Soybeans are a well-studied and widely applied as a source for protein in poultry and livestock diets. However, in recent years, concerns have mounted regarding the cultivation of soybean—particularly in topics such as world market power and the sustainability of production. Therefore, research institutions and industry alike are looking into the possible alternative animal feed protein sources available. The European Union (EU) relies on soybean imports to feed domestic poultry and livestock, with approximately 40% of animal feed protein originating from soy imports [2]. In addition, not only is the EU heavily dependent on soy imports, but China dominates the world import market by consuming approximately 41% of total world soy exports [3]. Furthermore, questions persist regarding the sustainability of soybean cultivation, primarily in the southern hemisphere. Prudêncio da Silva et al. [4] find that although deforestation is decreasing in Brazil, the previously incurred land use changes have led to secondary impacts, such as climate change and increased cumulative energy demand. Therefore, as a net importer of soy, the EU is conscious of their economic vulnerability and environmental responsibility.

This pilot study, focusing on poultry meat quality, is part of a larger project aimed at determining whether alternatives to soy can be incorporated into German meat production systems without having a negative effect on the overall product quality. Alternatives are deemed necessary in order to off-set soybean imports and decrease the European protein gap. Therefore, this study investigates the effects of two proteins that could be produced outside the arable farming system, focusing on one micro-algae source and one insect protein source, as exemplars. Spirulina (*Arthrospira platensis*) was chosen to partially replace soy because of its high crude protein content of 63% in dry matter (DM) [5] and the ability to cultivate it in photobioreactors or race-way ponds [2], which could reduce or at least limit the area required for cultivation in Europe. In addition, Spirulina contains antioxidants, such as β-carotene and vitamin E [6], which could have positive effects on meat physico-chemical parameters. *Hermetia illucens* L., also known as the black soldier fly, was also chosen as a prospective protein source for poultry diets because larvae also contain high amounts of crude protein [7] and could be fattened on numerous substrates, such as manure, cereals, and agro-food sector wastes [7,8], all of which are easily accessible in Europe, but are strictly regulated by the EU commission [9]. The conversion (recycling) of manure into a high quality animal feed [8] could be a notable advantage for Hermetia meal production; however, this is unlikely to be permitted within the EU. Unfortunately, to date in the EU, the use of animal originating feedstuffs is mostly prohibited in livestock production; nonetheless, insect feed is currently allowed in pet food and as of July 2017 has been approved for fish feed. Therefore, with continued legislative changes, Hermetia meal has the potential to be incorporated into poultry diets.

Although environmental and production sustainability are critical to ensuring the world's food supply, it should not have to come at an expense to quality. Therefore, we endeavour to capture the multi-faceted concept of meat quality through physico-chemical and sensory testing in order to determine the effects of off-setting soy with Spirulina or Hermetia meal. Ultimately, meat quality is a combination of factors [10] that differ from region to region [11] and in context [12]. Physico-chemical characteristics are usually instrumentally evaluated using literature-agreed-upon methods and parameters [13]. Whereas, sensory testing is the only way to fully quantify the aroma and texture characteristics experienced, often in combination, by consumers. This more complex testing is necessary because, of course, the palatability of meat is a combination of factors that cannot (yet) be captured simultaneously by laboratory techniques. Finally, in order to evaluate whether the alternatively-fed products can be immediately integrated into current production chains in Europe, primarily Germany, meat quality is assessed with samples stored in the industry's commonly practiced packaging [14]—highly oxygenated modified atmosphere packaging (HiOx MAP). The packaging type, length of storage time, and feed source can influence the flavour associated with aging or colour stability; therefore, in this pilot study, we investigate the impacts to physico-chemcial parameters and sensory properties when 50% of dietary soy protein is replaced by either *Arthrospira platensis* (Spirulina) or defatted dried *Hermetia illucens* larval meal (*Hermetia*) under current industrial packaging practices.

2. Materials and Methods

2.1. Animals and Diets

For this study, 132 Ross 308 male birds were raised on amino acid balanced diets where 50% of the soy-based protein was substituted by either Spirulina powder (n = 48) or Hermetia partially defatted larval meal (n = 48) in starter and grower diets; in the control group (n = 36) no soybean meal was substituted. The spirulina powder was sourced from Myanmar and had a moisture content of 3.4%. It was made up of 58.8% crude protein (DM) and 4.3% lipids (DM). The Hermetia meal was produced in Germany and the composition included 5.5% moisture, 60.5% crude protein (DM), and 14.1% lipids (DM). The main ingredients of the control diet during the starter, and respective (resp.) grower, period were wheat (32.9 resp. 37.6%), corn (16.4 resp. 18.8%), and soybean meal (39 resp. 32%). The diets were with amino acids supplemented according to the ideal amino acid ratio [15] and formulated to meet the energy and nutrient requirements of fast growing meat-type chickens according to current recommendations. Lysine level of the control diet was 1.25% and was 1.05% in the starter and grower periods. Table 1 outlines the diet composition for both starter and grower periods and the analysed diet nutritional content is listed in Table 2. The diets were created in reference to the Ross 308 nutritional specifications [16]. Further details to the procurement of Spirulina and Hermetia meal can be found in Neumann et al. [17].

Table 1. Ingredient composition of experimental diets (g/kg as fed).

Ingredients/Diets	Starter Period (1–21 day(s))			Grower Period (22–34 days)		
	Hermetia	Spirulina	Control	Hermetia	Spirulina	Control
Wheat	358.3	377.9	328.8	402.6	416.8	375.8
Corn	179.2	189.0	164.4	201.3	208.4	187.9
Soybean meal	195.0	195.0	390.0	160.0	160.0	320.0
Insect meal	145.4	-	-	119.0	-	-
Algae meal		118.2			97.0	
Soybean oil	78.5	78.5	78.5	78.5	78.5	78.5
Premix *	10.0	10.0	10.0	10.0	10.0	10.0
$CaHPO_4$	12.0	12.0	11.0	8.0	10.0	10.0
$CaCO_3$	9.9	9.1	11.0	8.0	8.0	9.0
NaCl	1.7	1.7	3.0	2.0	2.0	3.0
Wheat starch	-	-	-	3.0	3.0	-
TiO_2	-	-	-			3.0
L-Lysine·HCl	3.2	4.4	1.3	2.4	3.5	0.8
DL-Methionine	4.1	3.5	2.0	3.0	2.5	2.0
L-Threonine	0.6	-	-	0.4	-	-
L-Arginine	2.2	0.7	-	1.4	0.1	-
L-Valine	-	-	-	0.5	0.2	-

* Added per kg of final diet: 2.1 g calcium, 0.8 g sodium, 5000 IU vitamin A, 1000 IU vitamin D3, 30 mg vitamin E, 2.6 mg vitamin B1, 4.8 mg vitamin B2, 3.2 mg vitamin B6, 20 µg vitamin B12, 3 mg vitamin K3, 50 mg nicotinic acid, 10 mg calcium pantothenate, 0.9 mg folic acid, 100 µg biotin, 1000 mg choline chloride, 50 mg Fe as iron-II-sulfate, monohydrate, 15 mg Cu as copper-II-sulfate, pentahydrate, 120 mg Mn as manganese-II-oxide, 70 mg Zn as zinc oxide, 1.4 mg I as calcium iodate, hexahydrate, 0.28 mg Se as sodium selenite, 0.55 mg Co as alkaline cobalt-II-carbonate, monohydrate, and 100 mg butylhydroxytoluol.

Table 2. Analysed nutrient content of experimental diets (crude nutrients g/kg dry matter (DM)).

Diets	Starter Period (1–21 day(s))			Grower Period (22–34 days)		
	Hermetia	Spirulina	Control	Hermetia	Spirulina	Control
Crude protein	259.3	241.4	249.5	230.9	207.2	220.2
Ether extract	131.1	116.6	111.6	131.4	118.4	112.8
Crude fibre	47.1	31.1	45.2	41.7	30.4	40.4
Crude ash	60.4	59.2	65.6	56.5	53.5	61.6
N-free extract	502.1	551.7	528.1	539.5	590.5	565
AME_N (MJ/kg DM) *	15.3	15.4	14.4	15.6	15.6	14.8

* N corrected apparent metabolizable energy, calculated according to WPSA [18].

The diets were fed ad libitum and animals had constant access to water. The animals were standardly kept according to article 4 of Germany's Animal Welfare Regulation [19] on wood shaving covered floor pens (6 birds per pen; stocking density 5 birds per m^2) at the Division Animal Nutrition Physiology, University of Göttingen, Germany. In total, 8 pens per Spirulina or Hermetia treatment group were housed, and 6 pens were fed the control diet. Animals were randomized amongst the pens prior to slaughter to reduce a possible housing effect later in the experimental design. At 35 days of age, the animals were slaughtered by certified personnel at the University of Göttingen poultry slaughterhouse, which is authorized according to article 4 of the European Union's (EG) NR. 853/2004 [20]. Immediately following the slaughter, the *M. pectoralis superficialis* muscle (breast filet) was removed (excluding the *M. pectoralis profundus* muscle), and was cooled to 4 °C (ca. 5 h) until further processing, with the exception that pH and lean colour were monitored prior to cooling (see below).

2.2. Animal and Sample Management

In addition to the three feed treatment groups, the animals were further allocated into groups for physico-chemical or sensory testing, as can be seen in Figure 1. The first 12 carcasses per group (24 breast filets) were assigned for sensory testing and divided into 3 storage times so that 4 animals per feed treatment and storage time could be tested. The storage times were Fresh (not HiOx MAP packaged), 3 days, and 7 days. Two chicken filets derived from the same animal were packaged in polypropylene (PP) plastic trays (227/178/40 mm in dimension) lined with a moisture absorbent pad, heat-sealed with an oriented polyethylene terephthalate (OPET)/polypropylene (PP) film (<3 cm^3/m^2 24 h bar oxygen transmission rate; <12 cm^3/m^2 24 h bar carbon dioxide transmission rate; Dieter Seegers Haus der Verpackung GmbH, Osnabrück, Germany) and stored in an 80% O_2/20% CO_2 atmosphere. The packages were stored at 4 °C without illumination for the allotted time. Prior to sensory testing, the samples were vacuum-sealed in polyamide (PA)/polyethylene (PE) bags and frozen at −20 °C until further testing.

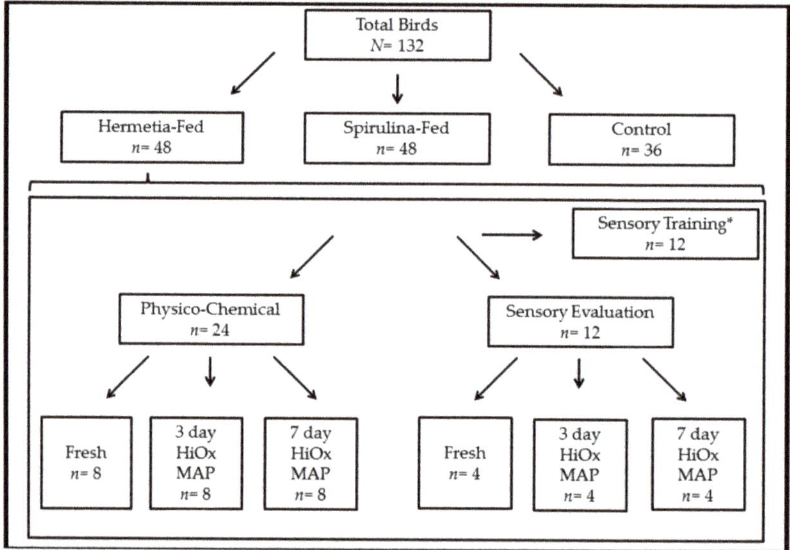

Figure 1. Allocation of birds (upper portion) and material within a treatment group (lower portion).
* Only for Spirulina- and Hermetia-fed groups

The next 24 carcasses per group were assigned for analytical meat quality testing and were also subsequently divided into 3 storage times: Fresh (n = 8), 3 days (n = 8), and 7 days (n = 8). The samples were packaged and stored as mentioned above. The following parameters were monitored: lean colour (20 min and 3 day or 7 day), pH (20 min and ultimate), lipid oxidation, storage loss (excluding Fresh samples), cooking loss, and shear force. The lipid oxidation samples were frozen at −70 °C prior to analysis and the right breast filet was frozen at −20 °C prior to cooking loss and subsequent shear force analysis. The additional animals in the Spirulina-fed (n = 12) and Hermetia-fed (n = 12) treatment groups (indicated with * in Figure 1) were also probed for pH_{20min} and pH_{24h} in the left filet, and the right filets were allocated for sensory training.

2.3. Physico-Chemical Characteristics

The pH was measured at 20 min post mortem and an ultimate pH measurement (pH_u) was taken at 24 h for the Fresh samples, and at 3 days or 7 days for the respective HiOx MAP samples. The measurements were taken with a portable pH meter equipped with a glass electrode and metal thermometer electrode (Knick Portamess 911, Berlin, Germany). The pH meter was calibrated every session prior to use using commercial pH 4 and 7 buffer solution standards (Merck, Darmstadt, Germany) at room temperature. The pH was measured by inserting the electrode completely into the superior portion of the left breast filet muscle. An accompanying thermometer was inserted alongside the electrode approximately 1 cm away. The lean colour was measured at 20 min post mortem and at 3 days or 7 days immediately after opening the package; no blooming time was allowed so that colour was recorded as close to consumers' perception through the packaging. CIELAB coordinate measurements were recorded and the values used in analysis were derived across the average of three measurements taken on the ventral side of the right breast filet using a portable spectrophotometer (model: CM 600d, Konica Minolta, Tokyo, Japan) with diffused illumination, an 8° viewing angle, and a silicon photodiode array as the detector. The spectrophotometer was calibrated using a white and black prop provided by the manufacturer prior to every session. In order to determine if the protein feed had an effect on meat composition, or whether meat composition could be significantly correlated with shear force values, breast filets were cleaned of excess subcutaneous fat, homogenized, and meat composition parameters were analysed using a Foss FoodScan™ according to Anderson [21]. To determine storage loss, the breast filets allocated for HiOx MAP were weighed prior to packaging and immediately after being removed from the package. Storage loss was expressed as the percent of weight loss over time compared to the initial sample weight. To determine cooking loss, the right breast filet, trimmed of exterior fat, was cooked sous vide for 60 min. The samples were vacuum-sealed in PA/PE bags and placed in a pre-heated hot water bath instrument (incubation/deactivation bath, Gesellschaft für Labortechnik mbH (GFL), Burgwedel, Germany) set to 77 °C. Samples were kept submerged and separated during cooking. Within the 60 min, the samples achieved 75 °C core temperatures, as determined with a preliminary study. After cooling to room temperature, the samples were weighed and cooking loss was expressed as the percentage of the initial weight. After determining cooking loss, the samples were stored at 4 °C for 24 h prior to shear force measurements. Shear force was determined according to Xiong et al. [22] with the following modifications: a TA.XTplus Texture Analyser (Stable Micro Systems, Surrey, UK) was set to a penetration depth of 15 mm to accommodate for thinner samples and a 50 N load cell was used. Each sample was tested 3 times and statistical analysis was conducted with the mean value across the 3 measurements. Shear force was measured as the highest peak (N).

2.4. Lipid Oxidation

In order to determine the extent of lipid oxidation over time, the 2-thiobarbituric acid reactive substances (TBARS) method was conducted according to Bruna et al. [23]. Results were recorded in terms of µg of malonaldehyde (MDA)/g of sample.

2.5. Sensory Analysis

Conventional profiling was used to determine the appropriate attributes to be later evaluated. Appropriate attributes were determined prior to evaluation by the assessors. In eight two-hour training sessions, the panel legitimated and defined the attributes unique amongst the products. These 19 attributes were used to evaluate the products in appearance, odour, taste and flavour, texture, and aftertaste. Overall odour intensity, animal or barn odour, metallic odour, cooked chicken odour, colour intensity (light-dark), visual elasticity assessment, fibrous appearance, overall flavour, sweet taste, sour taste, bitterness, metallic flavour, chicken flavour, overall aftertaste, hardness, juiciness, tenderness, adhesiveness, and crumbliness were evaluated (see Appendix A for more information).

The samples were cooked sous vide exactly as described for cooking loss above. The core temperature of every breast filet was checked to ensure that every sample was ca. 75 °C using a Testo 926 digital probe thermometer (Testo SE & Co. KGaA, Lenzkirch, Germany). The breast filets were then cut into 1 cm by 1cm pieces. Samples were served immediately on warmed plates and assigned with a 3 digit randomly allocated code. The evaluations took place in the University of Göttingen sensory laboratory, which is in compliance with international standard-ISO 8589 [24].

The trained panel consisted of ten assessors, who voluntarily provided written informed consent and were selected and trained according to ISO 8586-1 [25]. The panel evaluated chicken breast filet products differing in protein feed type (Spirulina, Hermetia, Control) and storage time (Fresh, 3 day HiOx MAP, 7 day HiOx MAP). In total, the nine products, derived from the three-by-three design, were evaluated in duplicate by each assessor over three one-hour sessions, where each assessor evaluated one sample for six of the nine products per session. The assessors evaluated the samples in a sequential monadic manner following four set orders that were randomly allocated, and a maximum of three assessors received the same set order during one session. The measurements were recorded electronically on 9 cm unmarked scales, a point system from 0 to 100 was stored behind the scale for later statistical analysis, using EyeQuestion survey software (Logic8 BV, Elst, The Netherlands). No assessor evaluated two samples from the same bird.

2.6. Statistical Analyses

The statistical analyses were carried out using SPSS (Version 24.0, IBM Corporation, Armonk, NY, USA) statistical software. One-way ANOVAs were conducted to determine the effect of feed on the carcass and meat quality parameters, such as live weight, carcass weight, the breast filet yield per animal, meat composition and pH_{20min} and pH_{24h}. Factorial ANOVA was carried out to determine the effects that feed, storage time (main effects), and a possible interaction effect (Feed × Storage) had on lean colour, lipid oxidation, storage loss, cooking loss, and shear force. The sensory data were evaluated using a linear mixed model with the sensory attributes as the dependent variables, feed, storage time in HiOx MAP, and a Feed × Storage interaction term as the fixed effects, and the random effects were listed as assessor and animal. Post Hoc tests were conducted using the Fischer's least significant difference (LSD) technique. Significant differences were determined where $p < 0.05$. In addition, effect size was calculated to account for possible Type 2 statistical errors, which could be likely due to the experiment's small sample size; by reporting the effect size one would nonetheless be able to compare our results to other studies and to interpret the relevance of effects as opposed to significance with the latter being substantially affected by sample size (increasing n ultimately leads to significance, even though the effect size is equally small as with a low n). Either Cohen's d for carcass performance parameters, where either Spirulina or Hermetia treatment group results were compared with the control, or partial eta-squared were calculated to estimate the effect size for physico-chemical data. Effect size was interpreted according to Cohen [26], where a small size is deemed to be 0.2 > 0.5, a moderate effect size is 0.8 < 0.5, and a large effect size is above 0.8. These values also correspond with the interpretation applied by Batorek et al. [27] in order to ascertain that the parametrical differences are derived from the treatments and not from zootechnical parameters, such as breast filet weight. Further, to maintain certainty that the significant differences were derived from the treatments, Pearson's r was also calculated between the weight of one breast filet and the corresponding dependent parameters—lean colour (L^*, a^*, b^*), lipid oxidation (TBARS), storage loss, cooking loss, and shear force, and between shear force values for Fresh samples and intramuscular fat (IMF) or moisture as dependent factors. A correlation was considered significant at a level of 0.05 using a two-tailed test.

3. Results

3.1. Physico-Chemical Results

There are only minor differences in carcass performance indicators between the feed treatment groups. Hermetia-fed birds were larger in size, which is accounted for by the statistically significant differences described in Table 3 for live weight and carcass weight. However, this difference in weight did not result in an overall increase in breast filet yield. Furthermore, the pH_{20min} and pH_{24h} values are significantly different between the different feed groups, with the control group having the highest pH values directly after slaughter, but the Spirulina-fed samples maintain a higher pH after 24 h.

Table 3. Means and standard deviation (SD) with Cohen's *d* for carcass performance parameters: live weight, carcass weight, breast filet yield, protein, intramuscular fat (IMF), moisture, and pH values at 20 min and 24 h post mortem.

Parameters	Hermetia	SD	Cohen's *d*	Spirulina	SD	Cohen's *d*	Control	SD
Live weight (g)	2329 ^{a} (n = 48)	322	0.509	2121 ^{b} (n = 48)	218	0.181	2169 ^{b} (n = 36)	306
Carcass weight (g)	1789 ^{a} (n = 48)	261	0.476	1577 ^{b} (n = 48)	175	0.465	1672 ^{b} (n = 36)	230
Breast filet yield (%)	20.4 ^{a} (n = 36)	1.9	0.205	20.0 ^{a} (n = 36)	2.0	0.400	20.8 ^{a} (n = 24)	2.0
Protein (% breast filet)	21.07 ^{a} (n = 8)	0.27	0.99	21.60 ^{a} (n = 8)	0.64	1.56	20.52 ^{a} (n = 8)	0.74
IMF (% breast filet)	3.41 ^{a} (n = 8)	0.45	0.02	3.12 ^{a} (n = 8)	0.50	0.47	3.40 ^{a} (n = 8)	0.68
Moisture (% breast filet)	73.88 ^{a} (n = 8)	0.43	0.62	73.62 ^{a} (n = 8)	0.71	0.87	74.29 ^{a} (n = 8)	0.83
pH_{20min}	6.65 ^{b} (n = 36)	0.12	0.917	6.67 ^{b} (n = 36)	0.12	0.750	6.76 ^{a} (n = 24)	0.12
pH_{24h}	5.80 ^{a} (n = 20)	0.18	0.00	6.06 ^{b} (n = 19)	0.13	1.852	5.80 ^{a} (n = 8)	0.15

Sample size indicated in brackets; superscript letter a–b indicate statistical differences between feed groups ($p < 0.05$); Cohen's *d* comparisons are between the respective treatment group and control group.

Regarding the breast filet quality, the estimated marginal means for the analysed quality characteristics are listed in Table 4. In terms of feed, nearly all differences are accounted for by the Spirulina feed. These samples were darker (L^* value), redder (a^* values), and more yellow (b^* values) in colour compared to the Hermetia-fed and control groups. The Spirulina-fed samples lost 0.5% less moisture while being stored and were able to retain that water even after cooking, as is shown by the nearly 2% decrease in cooking loss values compared to the Hermetia-fed group and 4% decrease in values compared to the control group. The Hermetia-fed values remained similar to those of the control group across all the studied parameters. TBARS and shear force values were not significantly different across the feed groups. In addition, breast filet weight (g) correlated minimally only with the lean colour parameters L^* ($r = 0.589$ with $p < 0.01$) and b^* ($r = 0.319$ with $p < 0.05$).

Table 4. Estimated marginal means with standard error (SE) and effect size denoted by partial η^2 for colour, lipid oxidation (2-thiobarbituric acid reactive substances (TBARS)), storage loss, cooking loss, and shear force.

Parameters	Feed Groups			Partial η^2	Storage Time			Partial η^2	SE
	Hermetia	Spirulina	Control		Fresh	3 Day	7 Day		
L^*	55.60 ^{a}	53.01 ^{b}	55.40 ^{a}	0.243	51.10 ^{f}	54.92 ^{e}	58.00 ^{d}	0.648	0.454
a^*	2.00 ^{b}	3.48 ^{a}	2.43 ^{b}	0.212	0.85 ^{f}	4.04 ^{d}	3.03 ^{e}	0.554	0.261
b^*	11.15 ^{b}	12.31 ^{a}	11.33 ^{b}	0.131	8.65 ^{e}	13.15 ^{d}	12.99 ^{d}	0.716	0.286
TBARS (μg/g)	0.106 ^{a}	0.095 ^{a}	0.098 ^{a}	0.006	0.081 ^{e}	0.084 ^{e}	0.134 ^{d}	0.150	0.013
Storage loss (%)	3.00 ^{a}	2.48 ^{b}	3.04 ^{a}	0.147	-	2.72 ^{d}	2.96 ^{d}	0.035	0.134
Cooking loss (%)	31.80 ^{a}	29.00 ^{b}	33.05 ^{a}	0.169	33.29 ^{d}	30.29 ^{e}	30.25 ^{e}	0.125	0.821
Shear force (N)	10.86 ^{a}	11.16 ^{a}	11.80 ^{a}	0.060	11.01 ^{d}	11.58 ^{d}	11.21 ^{d}	0.058	0.336

n = 24, except for storage loss (n = 16); superscript letter a–b indicate statistical differences between the feed groups ($p < 0.05$); superscript letter d–f indicate statistical differences between storage times ($p < 0.05$).

Concerning the effect of HiOx MAP storage, the results are not always so clear-cut. To start, the storage in HiOx MAP has an effect on the lightness values (L^*) with a steady increase in the values (lightness) over the three groups. HiOx MAP also appears to increase the a^* and b^* values; however, these values fall slightly between day 3 and day 7. The TBARS values are not significantly different between the Fresh samples and the 3 day HiOx MAP samples; however, the values do significantly increase for 7 day HiOx MAP samples. The HiOx MAP samples were significantly different from the Fresh samples in terms of cooking loss. The HiOx MAP samples lost 3% less moisture during cooking than their Fresh counterparts. Finally, the storage losses where similar between the HiOx MAP groups, despite the four day difference in storage times, and the shear force values were similar across the groups and not significantly correlated to meat composition parameters—IMF and moisture content.

Interestingly, the only significant interaction effect (Feed × Storage) is for pH_u ($p = 0.002$). Here we can see that after the initial fall of the pH_{20min} values to the pH_{24h} values (see Table 1), the values climb back up again over time in HiOx MAP (illustrated in Figure 2). The values are the highest for the control group compared to the two treatment groups, yet remain relatively stable between the 3 and 7 day tests for all HiOx MAP samples.

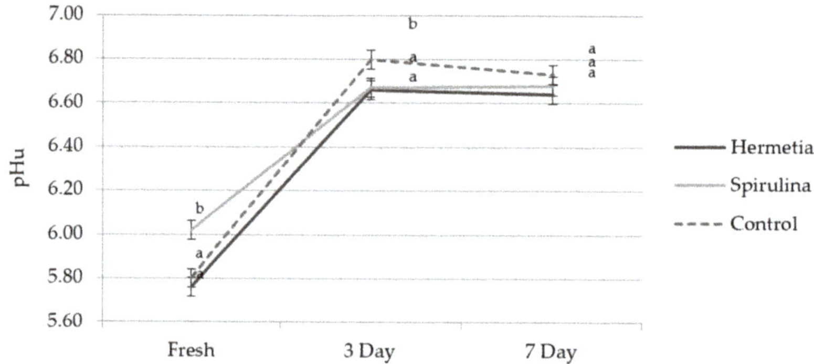

Figure 2. Interaction effect of feed and storage times ($p = 0.002$) on pH_u (mean, SE); superscript letter a–b denote statistical differences between the feed groups.

3.2. Sensory Results

Most attributes were not discernibly differentiated by the trained panel. However, the assessors determined that the feed significantly affected the hardness ($p < 0.003$) and tenderness ($p < 0.008$); storage time significantly affected elasticity when samples were stored up to 7 days ($p < 0.024$), and the interaction effect Feed × Time ($p < 0.002$) was significant for the overall flavour intensity. The effect of feed on metallic flavour was nearly significant (p-value = 0.051). Please refer to Table 5 to see the quantitative differences discerned. Generally, the Spirulina-fed samples were less metallic in flavour and the two alternative feed groups were softer and more tender than the control group.

Table 5. Estimated means with standard error (SE) and effect size (partial η^2) for the statistically significant sensory attributes (metallic flavour, hardness, and tenderness) according to feed type.

Sensory Attribute	Hermetia	Spirulina	Control	Partial η^2	SE
Metallic flavour (not metallic to strongly metallic)	20.4 ^{a}	15.8 ^{b}	17.1 ^{a}	0.223	3.90
Hardness (soft to hard)	28.8 ^{b}	25.0 ^{b}	40.8 ^{a}	0.300	4.93
Tenderness (tender to tough)	30.8 ^{b}	24.4 ^{b}	47.1 ^{a}	0.371	6.05

The table values are based on the 9 cm scale which was quantified from 0 to 100; superscript letter a–b denote statistical differences between the feed groups.

The interaction between feed type and HiOx MAP storage times is significant, and as illustrated in Figure 3 ($p = 0.002$), storage appears to have different effects depending on the feed type. For example: HiOx MAP storage appears to decrease the flavour intensity over time for Hermetia-fed samples, but not for Spirulina-fed or control samples. However, for all three groups, the 3 day HiOx MAP samples are less intense than the Fresh samples, and for the Spirulina-fed and control samples, the intensity climbs again after four additional days of aging in HiOX MAP. Finally, 7 day HiOx MAP (estimated mean of 34.0) results in a product that is less elastic compared to the Fresh (estimated mean of 41.9) and 3 day HiOx MAP (estimated mean of 41.3) breast filets ($p = 0.024$; SE= 5.41).

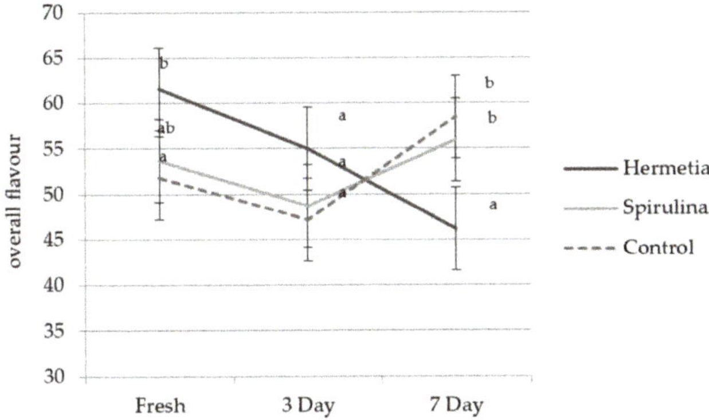

Figure 3. Interaction effect between feed and storage time on overall flavour ($p = 0.002$); superscript letter a–b denote statistical differences between the feed groups.

4. Discussion

The European protein gap is one reason why alternative protein sources should be widely studied [28]. Therefore, understanding the effects on meat quality from soybean meal substitution in poultry diets is important when considering what this elemental change could mean for packers, retailers, and consumers down the supply chain. At a quick glance, our results for the 50% substitution of soy-based protein with Spirulina or Hermetia larval meal, show very modest or no changes in the meat quality for many of the relevant parameters. The lack of correlation between breast filet size and the physico-chemical parameters shows that the significant differences listed above are likely not due to zootechnical differences in size between the groups, but rather due to the feed type and storage length treatments. The exceptions are L^* and b^*, where the larger filet size could influence the measureable differences; although if the filet size was the main influencing factor, then one would expect that the Hermetia-fed would be the diverging group, not the Spirulina-fed. The effect sizes, according to Cohen's d, imply that diet has a moderate effect on the carcass performance parameters, such as carcass weight and breast yield; however, the diet has a much larger effect on the pH value, especially for the Spirulina treatment group after 24 h, and type 2 statistical errors are likely not a factor with which to be concerned. According to the partial eta-squared parameters, it could be said that although feed may have a modest effect on meat quality, the type of packaging in conjunction with length of storage, in the end, plays a larger role in influencing meat quality. Although most of the results may be negligible, some of the deviations from the standard soy-fed control group should be well considered (lean colour) and appreciated (carcass weight) prior to fully incorporating alternative protein sources into poultry diets.

4.1. Spirulina in Poultry Diets

The incorporation of Spirulina results in a breast filet with a higher pH value at 24 h post mortem. In addition, storage and cooking losses decreased compared to the other two treatment groups, and this is likely related to the higher pH value. Although in our study the diets were only supplemented to balance dietary amino acids, increased dietary Lysine levels in diets has also resulted in similar findings of increased pH values and decreased storage losses [29]. Nonetheless, the result is improved meat quality, with a large effect size as according to the Cohen's d value. Reduced cooking loss should result in a breast filet that is expected to be 'juicier' [30] and more tender in texture [31]; however, sometimes these differences are first perceived by trained assessors when large differences of over 10% occur [32]. The trained panel in our study did not evaluate the Spirulina-fed breast filets as being juicier; however, Spirulina-fed breast filets are rated as being the most tender and the softest. Albeit, these traits are not significantly different as compared to those of the Hermetia-fed breast filets. Spirulina-fed breast filets also had the lowest values for metallic taste. Metallic flavour is usually considered an off-flavour in poultry products, such as white meat breast filets, given that it is a typical descriptor for game [33] and beef [34] products. Indeed, Brunton et al. [35] list high levels 1-octen-3-one as having a metallic odour associated with off-flavours in turkey breast meat, and Jayasena et al. [36] list metallic flavour as an off-flavour resulting from lipid oxidation in chicken meat. Therefore, reduced scores can be interpreted as having a positive impact on the flavour of Spirulina-fed breast filets.

The largest and most readily noticeable difference of Spirulina-fed breast filets is the significant intensive colour. Spirulina-fed breast filets are darker, redder, and more yellow in colour than the other two treatment groups tested in this study. These results are in alignment with Venkataraman et al. [37] and Toyomizu et al. [38], who noted the distinct colour of the breast filet when Spirulina was incorporated into poultry diets. The distinctive colour is not just apparent in instrumental spectrophotometer measurements, but also to the naked eye in the raw state when looking at the entire muscle. Surprisingly, when the breast filets are cooked, a colour difference between the products was not discernible by our trained panel, although we must note that it may be difficult to detect colour differences with only a 1 cm^2 surface area. The more intense colour is likely a result of the high amounts of carotenoids in Spirulina [6]. Holman and Malau-Aduli [39] believe this darker colour could be advantageous when feeding Spirulina to livestock, because meat colour is one of the most important quality indicators perceived by consumers [40–42]. However, precisely for this reason, consumer acceptance of and consumers' response to intensely pigmented poultry products needs to be studied, prior to incorporating Spirulina-fed products into the current production chain.

4.2. Hermetia in Poultry Diets

Hermetia larval meal could be used to substitute soybean meal, with minimal effects reaching the packers, retailers, or consumers. In fact, as with Spirulina, Hermetia feed could improve the overall meat quality. Hermetia animals were about 150 to 200 g heavier than their peers, and this translated into heavier dressed carcasses as well. Although the diets were calculated to be constant in caloric intake, the Hermetia meal diet was substantially higher in crude protein and ether extract (lipids), especially compared to the other treatment group, Spirulina. Therefore, it is likely that this effect is more due to the imbalances between the diets, than an inherent effect caused by Hermetia meal itself. The only other parameter that differed from the control group concerns the overall flavour intensity of Hermetia-fed breast filets. Here, a trend is noticed, where the Fresh Hermetia-fed breast filets score as the most intensive flavour; however, the intensity decreases when in HiOx MAP and over time. Although not fed with Hermetia, but rather house fly larvae, Gawaad and Brune [43] found that broiler chickens raised on (non-defatted) house fly and blow fly larvae meal had a unique smell and an intensive taste. However, this stronger taste does not need to be taken as a reduction in meat quality, because as Sheppard et al. [7] point out, some consumers prefer stronger tasting meat.

4.3. Impact of Storage Time

As previously mentioned, we packaged samples in HiOx MAP for up to 7 days, because in Germany, the country of study, this is commonly practiced by the industry for breast filet packaging and retailing [14]. We wanted to ensure that the alternatively-fed breast filets could be submitted to these conditions without reduced quality compared to the control product. In that regard, it appears that neither the Spirulina-fed nor the Hermetia-fed breast filets are negatively affected by HiOx MAP more so than the control. That being said, HiOx MAP storage over time did have some effects on overall product quality. For example, HiOx MAP increased the lightness, redness, and yellowness of the samples; although the samples were not able to maintain the complete colour change from day 3 until day 7. The values sank slightly over time. Given that the main reason for using HiOx MAP is to stimulate a more intensive red colour [42], this is not an unexpected result. Other effects include: increased TBARS values between the 3 day and 7 day HiOx MAP samples. This is expected granted that a HiOx environment induces lipid oxidation [44–48]. In addition, HiOx MAP breast filets had lower cooking losses compared to breast filets that were not packaged, but directly frozen to be analysed. This could be in part due to the additional storage time of 3 and 7 days, not specifically due to HiOx MAP. The additional aging time allowed the samples to lose about 3% of their weight; the difference in cooking loss between the HiOx MAP and Fresh breast filets is about 3%. The HiOx MAP breast filets did not lose more moisture if they were packaged for 7 compared to only 3 days. These results are similar to those of Delles and Xiong [44], who found limited changes in water-holding capacity between 4 and 14 days for HiOx MAP pork. It appears that there is a limit to storage loss amounts in HiOx MAP over time, *ceteris paribus*. Finally, 7 days of HiOx MAP impacted the elasticity of the cooked breast filets. The longer packaged breast filets were not as 'springy' when compressed with a fork and released to retake its original form. This could be due to protein oxidation [47] and a probable increase in myofibrillar deterioration [44].

4.4. Interaction Effects Feed × Storage

There are two significant interaction terms: (1) for ultimate pH and (2) for overall flavour intensity. First, the general increase in pH value in a HiOx MAP over time is supported by [43] who found that HiOx MAP increases pH values over time in pork products. However, the Spirulina- and Hermetia-fed breast filets appear to have more stable pH values in HiOx MAP compared to the control group. Although microbial indicators were not monitored in this study, according to Allen et al. [49], this could indicate that Hermetia- and Spirulina-fed breast filets could have a longer shelf-life. Secondly, the interaction term between the Hermetia-fed breast filets in HiOx MAP differs from that of the control breast filets. The Hermetia-fed breast filets taste less intense when packaged in HiOx MAP and the intensity decreases over the time that they are packaged. The Spirulina-fed and the control breast filets also decrease in flavour intensity when packaged, but the flavour intensity increases again from day 3 to day 7. This increase in intensity over time is expected granted that lipid oxidation increased between 3 and 7 days and lipid oxidation often leads to off-flavours in chicken meat [36].

4.5. Looking Forward

There are many parameters that should continue to be monitored in order to fully understand the impact that these alternative feeds have on the breast filet end product. The fatty acid composition and nutritive content, including vitamins, should be studied to determine if there are any nutritive benefits to the consumer. Studies should also monitor the pH stability of alternatively-fed products in turn with microbiological data to resolve whether these products could have longer shelf-lives than the standard fed product. In addition, focus should not be taken from determining the consumer response to products that potentially taste stronger or have a more intense colour at the point of purchase. Therefore, further research should include purchasing experiments based on poultry product colour, and more sensory testing with Hermetia-fed products should be executed. Despite the research gaps still present, we remain convinced that Hermetia and Spirulina are two good candidates to replace soybean meal in poultry diets.

5. Conclusions

Spirulina-fed breast filets generally have a higher pH, a higher water-holding capacity during storage and cooking, and have a reduced (metallic) off-flavour. The majority of the differences may be inconsequential; however, the intense colour of the Spirulina-fed breast filets should be further researched prior to incorporating Spirulina into poultry diets at a large scale. Hermetia-fed animals and carcasses are heavier and do not differ from the control group in other physico-chemical parameters. The breast filet has a more intense flavour when not HiOx packaged or aged over time, yet it remains to be seen if these differences are discernible by consumers. Based on their nominal effects on meat quality, Spirulina and Hermetia larval meal remain two potential protein alternatives for poultry diets.

Acknowledgments: This study is financed through the funding initiative *Niedersäschisches Vorab* by the Ministry for Science and Culture, Lower Saxony (#ZN 3041). Acknowledgments are also extended to Dieter Daniel for his expertise butcher skills, Christian Wagner for assisting in the slaughter organization, Ruth Wigger for her assistance with laboratory analysis, Claudia Kaltwasser for her support in the sensory laboratory, and Anina Vogt and Sylvia Lütke Daldrup for their support from slaughter to data entry. In addition, we would like to thank Janet Krickmeier and others from the Anhalt University of Applied Sciences for allowing us access to their Foss FoodScan^{TM} equipment expertise as well as Dieter Seegers Haus der Verpackung GmbH, Osnabrück for donating the packaging film used in this study.

Author Contributions: Daniel Mörlein and Frank Liebert conceived the study. Carmen Neumann and Susanne Velten designed the poultry diets and contributed their time and expertise raising the birds. Brianne A. Altmann performed the meat quality physico-chemical and sensory experiments. She also analysed the data and prepared the manuscript. All authors were involved in manuscript revisions and have read and approved the manuscript.

Conflicts of Interest: The authors declare no conflict of interest.

Appendix A.

Table A1. Sensory attributes identified and evaluated by trained assessors for characterizing chicken breast filets; scale from 0 to 100.

Attribute	Description	Sense	References *
1. Colour intensity	The intensity of the colour from white to dark beige	Visual	N/A
2. Elasticity	How far the sample can be pressed and still return to original form	Visual	<30: spreadable cheese 50: white bread 100: wine gum
3. Fibrousness	Fibre thickness on the product surface	Visual	<30: human hair <70: ca. 2 mm thick
4. Overall odour	The intensity of the smell when product held 2 cm from nose	Smell	N/A
5. Animal/Barn odour	Intensity of faecal, barn, or animal odour	Smell	Skatol
6. Metallic odour	Intensity of metal odour	Smell	old oxidized coins
7. Cooked chicken odour	Intensity of the odour from chicken meat in soup	Smell	cooked chicken juice
8. Overall flavour	Intensity of the overall flavour	Taste	N/A
9. Sweet taste	Intensity of sweetness	Taste	sucrose solutions 50: 6 g/L 100: 18 g/L
10. Sour taste	Intensity of sour taste	Taste	citric acid solutions 50: 0.28 g/L 100: 0.4 g/L
11. Bitter taste	Intensity of bitterness	Taste	caffeine solution 50: 0.21 g/L 100: 0.3 g/L
12. Metallic flavour	Intensity of metallic taste, such as serum or old coins	Taste	N/A
13. Chicken flavour	Intensity of taste similar to chicken soup	Taste	chicken soup broth
14. Aftertaste	The intensity of the aftertaste after swallowing	Taste	N/A
15. Hardness	The force needed to bite through the sample	Texture	<30: spreadable cheese 50: gouda cheese 100: Werther's Original
16. Juiciness	Amount of fluid released from product upon first bite	Texture	<30: raw carrot 50: cucumber 100: orange
17. Tenderness	The degree that the sample remains in one piece while chewing	Texture	<30: raw carrot 50: gouda cheese 100: hard candy
18. Adhesiveness	The degree the product sticks to your teeth; force needed to pull teeth apart	Texture	50: spreadable cheese 100: peanut butter
19. Crumbliness	Number of particles in mouth directly prior to swallowing	Texture	N/A

* Quantitative/qualitative.

References

1. OECD/Food and Agriculture Organization of the United Nations. *OECD-FAO Agricultural Outlook 2015–2024*; OECD/Food and Agriculture Organization of the United Nations: Rome, Italy, 2015. [CrossRef]
2. Taelman, S.E.; De Meester, S.; Van Dijk, W.; da Silva, V.; Dewulf, J. Environmental sustainability analysis of a protein-rich livestock feed ingredient in The Netherlands: Microalgae production versus soybean import. *Resour. Conserv. Recycl.* **2015**, *101*, 61–72. [CrossRef]
3. Song, B.; Marchant, M.A.; Reed, M.R.; Xu, S. Competitive analysis and market power of China's soybean import market. *Int. Food Agribus. Manag. Rev.* **2009**, *12*, 21–42.
4. Da Silva, V.P.; van der Werf, H.M.G.; Spies, A.; Soares, S.R. Variability in environmental impacts of Brazilian soybean according to crop production and transport scenarios. *J. Environ. Manag.* **2010**, *91*, 1831–1839. [CrossRef] [PubMed]
5. Tokusoglu, Ö.; Unal, M.K. Biomass nutrient profiles of three microalgae: *Spirulina platensis*, *Chlorella vulgaris* and *Isochrysis galbana*. *J. Food Sci.* **2003**, *68*, 1144–1148. [CrossRef]
6. Habib, M.A.B.; Parvin, M.; Huntington, T.C.; Hasan, M.R. A Review on Culture, Production and Use of Spirulina as Food for Humans and Feeds for Domestic Animals and Fish. In *FAO Fisheries and Aquaculture Circular*; FAO: Rome, Italy, 2008; No. 1034; ISBN 978–92-5-106106-0.
7. Spranghers, T.; Ottoboni, M.; Klootwijk, C.; Ovyn, A.; Deboosere, S.; De Meulenaer, B.; Michiels, J.; Eeckhout, M.; De Clercq, P.; De Smet, S. Nutritional composition of black soldier fly (*Hermetia illucens*) prepupae reared on different organic waste substrates. *J. Sci. Food Agric.* **2016**, *97*, 2594–2600. [CrossRef] [PubMed]
8. Newton, G.L.; Sheppard, D.C.; Burtle, G.J. Research Briefs: Black Soldier Fly Prepupae—A Compelling Alternative to Fish Meal and Fish Oil. Available online: http://www.extension.org/pages/15054/research-summary:-black-soldier-fly-prepupae-a-compelling-alternative-to-fish-meal-and-fish-oil (accessed on 8 August 2017).
9. European Union. Commission Regulation (EU) 2017/893. Available online: http://eur-lex.europa.eu/legal-content/EN/TXT/?uri=CELEX:32017R0893 (accessed on 6 November 2017).
10. Toldra, F.; Flores, M. Analysis of Meat Quality. In *Handbook of Food Analysis*, 2nd ed.; Nollet, L.M.L., Ed.; CRC Press: Boca Raton, FL, USA, 2004; Volume 3, pp. 1961–1978. ISBN 978-1-4398-3305-6.
11. Oliver, M.A.; Nute, G.R.; Font-i-Furnols, M.; San Julián, R.; Campo, M.M.; Sañudo, C.; Cañeque, V.; Guerrero, L.; Alvarez, I.; Díaz, M.T.; et al. Eating quality of beef, from different production systems, assessed by German, Spanish and British consumers. *Meat Sci.* **2006**, *74*, 435–442. [CrossRef] [PubMed]
12. Korzen, S.; Lassen, J. Meat in context. On the relation between perceptions and contexts. *Appetite* **2010**, *54*, 274–281. [CrossRef] [PubMed]
13. Font-i-Furnols, M.; Candek-Potokar, M.; Maltin, C.; Prevolnik Povše, M. *A Handbook of Reference Methods for Meat Quality Assessment*; European Cooperation in Science and Technology (COST): Brussels, Belgium, 2015.
14. BfR. Fragen und Antworten zu Fleisch, Welches unter Schutzatmosphäre mit Erhöhtem Sauerstoffgehalt Verpackt Wurde. Bundesinstitut für Risikobewertung. Available online: http://www.bfr.bund.de/de/fragen_und_antworten_zu_fleisch__welches_unter_schutzatmosphaere_mit_erhoehtem_sauerstoffgehalt_verpackt_wurde-51981.html#topic_51983 (accessed on 25 July 2017). (In German)
15. Wecke, C.; Liebert, F. Improving the reliability of optimal in-feed amino acid ratios based on individual amino acid efficiency data from N balance studies in growing chicken. *Animals* **2013**, *3*, 558–573. [CrossRef] [PubMed]
16. Ross 308 Broiler: Nutrition Specifications (All Plant-Protein Based Feeds). Available online: http://eu.aviagen.com/assets/Tech_Center/Ross_Broiler/Ross-308-Broiler-Nutrition-Specs-plant-2014-EN.pdf (accessed on 22 February 2018).
17. Neumann, C.; Velten, S.; Liebert, F. Improving the dietary protein quality by amino acid fortification with a high inclusion level of micro algae (*Spirulina platensis*) or insect meal (*Hermetia illucens*) in meat type chicken diets. *Open J. Anim. Sci.* **2018**, *8*, 12–26. [CrossRef]
18. WPSA. The Prediction of Apparent Metabolizable Energy Values for Poultry in Compound Feeds. *Worlds Poult. Sci J.* **1984**, *40*, 181–182.
19. Verordnung zum Schutz Landwirtschaftlicher Nutztiere und Anderer zur Erzeugung Tierischer Produkte Gehaltener Tiere bei ihrer Haltung. Available online: http://www.gesetze-im-internet.de/tierschnutztv/ (accessed on 6 August 2017). (In German)

20. European Communities. Regulation (EC) NO 853/2004 of the European Parliament and of the Council. Available online: http://eur-lex.europa.eu/legal-content/EN/TXT/?qid=1520433365249&uri=CELEX:32004R0853 (accessed on 6 August 2017).
21. Anderson, S. Determination of Fat, Moisture, and Protein in Meat and Meat Products by Using the FOSS FoodScan™ Near-Infrared Spectrophotometer with FOSS Artificial Neural Network Calibration Model and Associated Database: Collaborative Study. *J. AOAC Int.* **2007**, *90*, 1073–1083. [PubMed]
22. Xiong, R.; Cavitt, L.C.; Meullenet, J.F.; Owens, C.M. Comparison of Allo-Kramer, Warner-Bratzler and razor blade shears for predicting sensory tenderness of broiler breast meat. *J. Text. Stud.* **2006**, *37*, 179–199. [CrossRef]
23. Bruna, J.M.; Ordóñez, J.A.; Fernández, M.; Herranz, B.; De La Hoz, L. Microbial and physico-chemical changes during the ripening of dry fermented sausages superficially inoculated with or having added an intracellular cell-free extract of *Penicillium aurantiogriseum*. *Meat Sci.* **2001**, *59*, 87–96. [CrossRef]
24. International Organization for Standardization. ISO 8589:2007: Sensory Analysis—General Guidance for the Design of Test Rooms. Available online: https://www.iso.org/standard/36385.html (accessed on 20 January 2018).
25. International Organization for Standardization. ISO 8586-1:1993: Sensory Analysis—General Guidance for the Selection, Training and Monitoring of Assessors—Part 1: Selected Assessors. Available online: https://www.iso.org/standard/15875.html (assessed on 20 January 2018).
26. Cohen, J. The Effect Size Index: D. In *Statistical Power Analysis for the Behavioural Sciences*, 2nd ed.; Lawrence Erlbaum Associates: Mahwah, NJ, USA, 1988; pp. 20–27. ISBN 0805802835.
27. Batorek, N.; Čandek-Potokar, M.; Bonneau, M.; Van Milgen, J. Meta-analysis of the effect of immunocastration on production performance, reproductive organs and boar taint compounds in pigs. *Animal* **2012**, *6*, 1330–1338. [CrossRef] [PubMed]
28. Martin, N. What is the way forward for protein supply? The European perspective. *Oilseeds Fats Crops Lipids* **2014**, *21*, D403. [CrossRef]
29. Berri, C.; Besnard, J.; Relandeau, C. Increasing Dietary Lysine Increases Final pH and Decreases Drip Loss of Broiler Breast Meat. *Poult. Sci.* **2008**, *87*, 480–484. [CrossRef] [PubMed]
30. Aaslyng, M.D.; Bejerholm, C.; Ertbjerg, P.; Bertram, H.C.; Andersen, H.J. Cooking loss and juiciness of pork in relation to raw meat quality and cooking procedure. *Food Qual. Prefer.* **2003**, *14*, 277–288. [CrossRef]
31. Liu, Y.; Lyon, B.G.; Windham, W.R.; Lyon, C.E.; Savage, E.M. Principal component analysis of physical, color and sensory characteristics of chicken breasts deboned at two, four, six and twenty-four hours postmortem. *Poult. Sci.* **2004**, *83*, 101–108. [CrossRef] [PubMed]
32. Choi, Y.-S.; Hwang, K.E.; Jeong, T.J.; Kim, Y.B.; Jeon, K.H.; Kim, E.M.; Sung, J.M.; Kim, H.W.; Kim, C.J. Comparative study on the effects of boiling, steaming, grilling, microwaving and superheated steaming on quality characteristics of marinated chicken steak. *Korean J. Food Sci. Anim. Resour.* **2016**, *36*, 1–7. [CrossRef] [PubMed]
33. Neethling, J.; Hoffman, L.C.; Muller, M. Factors influencing the flavour of game meat: A review. *Meat Sci.* **2016**, *113*, 139–153. [CrossRef] [PubMed]
34. Campo, M.M.; Nute, G.R.; Hughes, S.I.; Enser, M.; Wood, J.D.; Richardson, R.I. Flavour perception of oxidation in beef. *Meat Sci.* **2006**, *72*, 303–311. [CrossRef] [PubMed]
35. Brunton, N.P.; Cronin, D.A.; Monahan, F.J.; Durcan, R. A comparison of solid-phase microextraction (SPME) fibres for measurement of hexanal and pentanal in cooked turkey. *Food Chem.* **2000**, *68*, 339–345. [CrossRef]
36. Jayasena, D.D.; Ahn, D.U.; Nam, K.C.; Jo, C. Flavour chemistry of chicken meat: A review. *Asian-Australas. J. Anim. Sci.* **2013**, *26*, 732–742. [CrossRef] [PubMed]
37. Toyomizu, M.; Sato, K.; Taroda, H.; Kato, T.; Akiba, Y. Effects of dietary Spirulina on meat colour in muscle of broiler chickens. *Br. Poult. Sci.* **2001**, *42*, 197–202. [CrossRef] [PubMed]
38. Venkataraman, L.V.; Somasekaran, T.; Becker, E.W. Replacement value of blue-green alga (*Spirulina platensis*) for fishmeal and a vitamin-mineral premix for broiler chicks. *Br. Poult. Sci.* **1994**, *35*, 373–381. [CrossRef] [PubMed]
39. Holman, B.W.B.; Malau-Aduli, A.E.O. Spirulina as a livestock supplement and animal feed. *J. Anim. Physiol. Anim. Nutr.* **2013**, *97*, 615–623. [CrossRef] [PubMed]
40. Viana, E.S.; Gomide, L.A.M.; Vanetti, M.C.D. Effect of modified atmospheres on microbiological, color and sensory properties of refrigerated pork. *Meat Sci.* **2005**, *71*, 696–705. [CrossRef] [PubMed]
41. McMillin, K.W. Where is MAP Going? A review and future potential of modified atmosphere packaging for meat. *Meat Sci.* **2008**, *80*, 43–65. [CrossRef] [PubMed]

42. Carpenter, C.E.; Cornforth, D.P.; Whittier, D. Consumer preferences for beef color and packaging did not affect eating satisfaction. *Meat Sci.* **2001**, *57*, 359–363. [CrossRef]
43. Gawaad, A.; Brune, H. Insect protein as a possible source of protein to poultry. *Z. Tierphysiol. Tierernaehrung Futtermittelkd* **1979**, *42*, 216–222. [CrossRef]
44. Delles, R.M.; Xiong, Y.L. The effect of protein oxidation on hydration and water-binding in pork packaged in an oxygen-enriched atmosphere. *Meat Sci.* **2014**, *97*, 181–188. [CrossRef] [PubMed]
45. Spanos, D.; Tørngren, M.A.; Christensen, M.; Baron, C.P. Effect of oxygen level on the oxidative stability of two different retail pork products stored using modified atmosphere packaging (MAP). *Meat Sci.* **2016**, *113*, 162–169. [CrossRef] [PubMed]
46. Bao, Y.; Puolanne, E.; Ertbjerg, P. Effect of oxygen concentration in modified atmosphere packaging on color and texture of beef patties cooked to different temperatures. *Meat Sci.* **2016**, *121*, 189–195. [CrossRef] [PubMed]
47. Xiong, Y.L.; Park, D.; Ooizumi, T. Variation in the Cross-Linking Pattern of Porcine Myofibrillar Protein Exposed to Three Oxidative Environments. *J. Agric. Food Chem.* **2008**, *57*, 153–159. [CrossRef] [PubMed]
48. Muhlisin, P.; Kim, D.S.; Song, Y.R.; Lee, S.J.; Lee, J.K.; Lee, S.K. Effects of gas composition in the modified atmosphere packaging on the shelf-life of *Longissimus dorsi* of Korean native black pigs-duroc crossbred during refrigerated storage. *Asian-Australas. J. Anim. Sci.* **2014**, *27*, 1157–1163. [CrossRef] [PubMed]
49. Allen, C.D.; Russell, S.M.; Fletcher, D.L. The relationship of broiler breast meat color and pH to shelf-life and odor development. *Poult. Sci.* **1997**, *76*, 1042–1046. [CrossRef] [PubMed]

MDPI
St. Alban-Anlage 66
4052 Basel
Switzerland
Tel. +41 61 683 77 34
Fax +41 61 302 89 18
www.mdpi.com

Foods Editorial Office
E-mail: foods@mdpi.com
www.mdpi.com/journal/foods

www.ingramcontent.com/pod-product-compliance
Lightning Source LLC
LaVergne TN
LVHW070703170726
843515LV00004B/473
* 9 7 8 3 0 3 9 4 3 3 7 2 8 *